Munching Maggots, Noah's Flood & TV Heart Attacks
and other cataclysmic science moments

GW00505992

Munching Maggots, Noah's Flood & TV Heart Attacks

and other cataclysmic science moments

Dr KARL KRUSZELNICKI'S

new moments in science # 3

Illustrations by Karen Young and Peter Pound

HarperCollins*Publishers*

HarperCollins*Publishers*

First published in Australia in 1998
by HarperCollins*Publishers* Pty Limited
ACN 009 913 517
A member of HarperCollins*Publishers* (Australia) Pty Limited Group
http://www.harpercollins.com.au

HarperCollins*Publishers*
25 Ryde Road, Pymble, Sydney, NSW 2073, Australia
31 View Road, Glenfield, Auckland 10, New Zealand
77–85 Fulham Palace Road, London W6 8JB, United Kingdom
Hazelton Lanes, 55 Avenue Road, Suite 2900, Toronto, Ontario M5R 3L2
and 1995 Markham Road, Scarborough, Ontario M1B 5M8, Canada
10 East 53rd Street, New York NY 10032, USA

National Library Cataloguing-in-Publication data:

Kruszelnicki, Karl, 1948– .
Munching maggots, Noah's flood and TV heart attacks.
Bibliography
ISBN 0 7322 5858 8.
1. Science–Popular works. I. Young, Karen, 1967– .
II. Pound, Peter. III. Title. (Series: New moments in science; 3).

500

Cover illustration by Karen Young
Cartoons by Karen Young
Technical illustrations by Peter Pound
Printed in Australia by Griffin Press Pty Ltd

5 4 3 2 1
02 01 00 99 98

THANKS

This book exists because of the work of many people. I do not do any research – I simply read the work of real scientists who do.

So first, I would like to thank the scientists who actually did the research work in the first place. (There are about 700,000 scientists in the USA alone!) Scientists like these have made our world a better place – such as giving us a lower Infant Mortality Rate (the number of babies that die in the first year of life, out of every thousand babies born alive).

Second, I thank my family (Mary, Little Karl, Alice and Lola) for listening to, and criticising, the stories, and making them better than they would have otherwise been.

Third, I thank the staff at HarperCollins Australia – Lore Foye (book designer), Gwenda Jarred and Amanda Berridge (legals), Kate Thomas (possibly the most effective publicity agent in the Entire Known Universe), Graeme Jones (typesetter), Kylie Corrigan (production controller) and Dr Belinda Lee (the editor, who even knows what a gerundive is!). Karen Young and Peter Pound did the great illustrations.

Fourth, I thank Dan Driscoll (ABC radio) for turning the short, rough radio stories into polished gems (before they changed into book stories), and Woo Wei for her comments and punchlines.

Fifth, I thank the people who made, and maintain, the more-than 750,000 words on our magnificent homepage (www.abc.net.au/science/k2) at The Lab – Ian Allen, Karen 'Cookie' Cook and Damon Shorter.

Finally, I thank Caroline Pegram for her magnificent research skills in being able to find any piece of information known to the human race (whether from CD-ROMs, the Net, or from written sources) – in minutes!

And I offer a free autographed copy of my next book to anybody who is the first to pick up a 'science' mistake in this book.

KARL KRUSZELNICKI

RUNNING IN THE RAIN

In the movie *Butch Cassidy and the Sundance Kid*, there's a famous song called 'Raindrops Keep Falling On My Head'. Unfortunately, raindrops also fall on the front and back and sides of your body — and you get wet. Usually, people prefer not to get wet in the rain, and run away from some raindrops that they would have otherwise caught. When most of us get caught in the rain without an umbrella, we make a mad *instinctive* dash for the nearest shelter.

But is our natural instinct to run, correct? *Scientifically*, what is the best tactic to use when you're caught in the rain — run, or walk?

It's actually quite a tricky question. If you run, you'll spend less time in the rain. On the other hand, you'll run into some raindrops that would have otherwise missed you. Which effect is greater?

Run or Walk Debate – 1975

One of the first people to write about this was Jearl Walker back in 1975, in his famous book *The Flying Circus of Physics*. Basically, he favoured running instead of walking — but he didn't really explain why.

Angle of Falling Rain

Jearl Walker mainly talked about the angle at which the rain is falling. If it's coming from directly above, or from in front of you, his advice is that you should run as quickly as possible to the nearest shelter. If the rain is falling on you from behind, you should try to run at the same horizontal speed as the rain. In other words, if you were running at the same horizontal speed as the falling rain, you would sweep up (or soak up) only one lot of raindrops — and you should stay fairly dry after that.

The 'run or walk-in-the-rain' question had been asked, but had not been fully answered.

Run or Walk Debate – January 1995

Nobody did much about the 'run or walk-in-the-rain' controversy for the next 20 years. But in January 1995, this same question appeared again in the back pages of the *New Scientist* magazine — along with three answers from three readers.

Wind Direction

Matthew Wright from the University of Southampton agreed with Jearl Walker

WATER IN THE AIR

If all the water vapour in the atmosphere were condensed into a single layer of water over the whole planet, that layer would be 25 mm (1 inch) thick.

Water molecules have a cycle where they evaporate from the surface, travel through the atmosphere, and then fall as precipitation (liquid water, ice, snow, etc.). The average rainfall is about 1000 mm (about 40 inches). So the average annual time that an individual molecule of water spends in the atmosphere is about nine days.

about the importance of wind direction. His answer was a limerick:

When caught in the rain without mac,
Walk as fast as the wind at your back,
But when the wind's in your face
The optimal pace
Is fast as your legs can make track.

Horizontal Rain

Martin Whittle wrote about the weather in the Lake District in England, where horizontal rain happens quite frequently. He agreed with Jearl Walker when he said that 'by running to keep up with the rain it is theoretically possible to stay dry'.

Evaporation

Mike Stevenson talked about your clothes drying due to evaporation, and said that if you ran quickly, you'd have less time for this evaporative drying. He recommended that if the rain was quite light, you should probably walk.

The readers each had an answer — *but none of them had done the experiment.*

Run or Walk Debate – November 1995

Now this kind of soft theorising from the back pages of the *New Scientist* wasn't good enough for some meteorologists from the University of Reading. So in November 1995, Holden, Belcher, Horvath and Pytharoulis (hereafter called HBHP) released their ground-breaking paper, 'Raindrops Keep Falling on my Head'.

Like good scientists, they first searched the literature for all of the other scientific papers that had been written on this important topic of whether to run or walk in the rain. They then used this knowledge as a base to do more research — by adding mathematics.

They realised that there are two ways in which a human can be made wet by a bunch of raindrops. First, the raindrops can fall on the top of your head. Second, as you move forward, you and your clothes can sweep up any raindrops that you happen to run into.

MAKING RAINDROPS

Making raindrops is a two-stage process.

In the first stage, as the water vapour rises from the oceans and the land, it cools. The water vapour will 'condense' on a tiny particle called a 'condensation nucleus' into tiny 'cloud droplets' of liquid. Condensation nuclei can include sea salt particles, combustion particles from fires, and clay-silicate particles lifted from the ground. These condensation nuclei are very small (0.002 mm), and need only very tiny upward wind currents (0.0000001 metres per second) to keep them floating. The cloud droplets (0.01 to 0.02 mm) are much bigger than the condensation nuclei, but are still small enough to be kept floating by a small upward wind current (0.01 m/s).

In the second stage, the cloud droplet has to grow 100 times bigger. One way to do this is by a process called 'collision and coalescence'. Slightly bigger droplets will fall through the other droplets. As they fall, they can sweep up other droplets, and so grow bigger. This process is very effective in clouds in the tropics, which have turbulent wind currents inside the cloud to bring droplets into contact with other droplets.

First Theoretical Model

They assumed that a person was basically a 3-D rectangle, with different areas on the top, front and sides. They then set up a few equations to model the movement of this 3-D rectangle through a space filled with air and raindrops, and then solved these equations with a high-powered mathematics called calculus. They also assumed that an average walking pace is about 2 to 3 metres per second (m/s), and looked at three different rainfall rates (5, 10 and 15 mm/hour). Finally, after plugging in these numbers and then drawing some graphs, they had their answer. They concluded that if you ran as quickly as you could, you would soak up 10 per cent less water than if you walked — hardly worth it! But they were wrong!

To Run or Not to Run?

They were a little surprised by this result. After all, in real life, people run in the rain, but their theoretical calculations did not show any big benefit in running.

But once again, the experiment had not been done.

Run or Walk Debate - 1997

In March 1997, two meteorologists, Thomas C. Peterson and Trevor W.R. Wallis, from the National Climatic Data Center in North Carolina in the USA, took the 'run/walk'

NO TEAR-SHAPED RAINDROPS

You'll often see a falling raindrop drawn by a cartoonist as being round at the bottom, with a pointed tip at the top. It isn't real!

Raindrops are quite close to round in shape, up to a size of about 2 mm. Above that size, as they get bigger, they get more flattened at the bottom. Once raindrops get to about 6 mm in diameter, they become unstable and can break apart due to the airflow.

debate a little further. They published a paper called 'Running in the Rain'.

These meteorologists wrote their paper because they had been soaked by rain while jogging through the southern Appalachian forest near their office. They weren't surprised by the rain because, as meteorologists, they had watched the weather forecast. But as the rain gradually became heavier, and they slowed down as they ran up a steep hill, they began to speculate about what was the best speed at which to travel.

Once they were back in the office, they read the 'Raindrops Keep Falling on my Head' paper by HBHP. They found a few minor mistakes, and corrected them.

MISTAKE No. 1: Peterson and Wallis realised that HBHP had overestimated the walking speed and running speed of humans.

HBHP had written that the average walking speed was about 2 to 3 m/s. HBHP probably got slightly confused by using scientific units (m/s), rather than commonplace units (such as kilometres or miles per hour). Two point five m/s works out to 9 kph — definitely faster than a walking pace. A typical walking pace is about 1.5 m/s (5.4 kph or 3.4 mph). An Olympic runner could just reach 8 m/s (28.8 kph, or 18 mph), so an average fit citizen in street clothing could probably reach 4 m/s (14.4 kph, or 9 mph) on a wet and slippery surface.

MISTAKE No. 2: Secondly, Peterson and Wallis found a mistake in the calculations of HBHP.

For example, look at the case of a person with a top surface area of 0.1 square metres and a cross-sectional area of 0.6 square

GUINNESS RAIN RECORDS

According to *The Guinness Book of Records*, the most intense rain-burst recorded in modern times was 38 mm (1.5 inches) in one minute at Barst in Guadeloupe on 26 November 1970.

Again according to *The Guinness Book of Records*, the location with more rainy days than anywhere else in the world is Mt Wai-'ale-'ale, on the island of Kauai in Hawaii — up to 350 rainy days per year!

Going by average yearly rainfall, the wettest place in the world is in India — Mawsynram, in the State of Meghalaya. It has 11.875 metres (467.5 inches) each year. In a single calendar month, the greatest rainfall ever recorded was 9.296 metres (366 inches), also in the State of Meghalaya in India, in July 1861, at Cherrapunji. Cherrapunji also holds the record for the greatest yearly rainfall — 26.461 metres (1041.8 inches) from August 1860 to July 1861.

And the greatest rainfall ever in a 24-hour period was recorded on 15 and 16 March 1952, in Cilaos at La Réunion in the Indian Ocean. The rainfall was measured at 1.87 metres, which works out to 3428 tonnes of water per hectare (8327 tons of water per acre).

metres. Suppose that this person has to travel 100 metres at 1 m/s (3.6 kph or 2.24 mph) through rain falling at 10 mm/hour. HBHP's calculations showed that this person would encounter 0.18 kg (180 grams, or 6.35 ounces, or oz) of water — but Peterson and Wallis used HBHP's own equations to come up with 0.06 kg (60 grams, or 2.13 oz). HBHP had made a simple mathematical mistake.

So when Peterson and Wallis used their more accurate assumptions (about walking and running speeds) and more accurate

SIZE OF DROPS

In a white fluffy cloud, the diameter of the 'cloud droplets' is about 0.01 to 0.02 mm. 'Drizzle drops' are defined to be between 0.2 and 0.5 mm. A typical raindrop is about 1 to 2 mm in diameter, while large raindrops from a thunderstorm can be between 5 and 8 mm in diameter.

There are two forces acting on raindrops — the 'suck' of gravity downwards, and the 'up force' of the updraught of the wind.

Cloud droplets are so small, that the wind speed needed to keep them aloft is very small — 0.01 m/s. But this wind speed needed increases as the size of the droplet increases — 0.7 m/s for a drizzle drop 0.2 mm in size, 4 m/s for a raindrop 1.0 mm in size, and 9 m/s for a large raindrop 5 mm in diameter.

calculations, they found that a walking person would encounter 0.052 kg (52 grams, 1.83 oz) of water, but that a running person would soak up 0.040 kg (40 grams, 1.41 oz) — a 23 per cent improvement. Twenty-three per cent is a lot better than the 10 per cent improvement of HBHP — and suddenly, running in the rain made more sense.

Second Theoretical Model

But then Peterson and Wallis went a little further. HBHP were one of the first to explore this topic, and quite clearly stated that their model was simple. Peterson and Wallis had the advantage of reading HBHP's work, and so they came up with a more sophisticated model that was closer to the truth.

They took into account another three factors — the velocity of the falling rain, the effect of leaning forward as you run, and rain that is driven by the wind.

Their improved model showed that in a light rain with no wind at all, running will give you only a 16 per cent reduction in wetness, as compared to walking. But if you're running rapidly and leaning forward in a heavy rain that is driven by the wind, you will end up 44 per cent less wet than if you had walked.

The Experiment

At this stage, Peterson and Wallis showed that they were real scientists, and decided to *do the experiment*.

They didn't need an $80 million satellite or complex lab equipment — which was just as well, seeing that they were paying for this out of their own pockets, and were doing the experiment on their own time. (They had done all their work because of 'intellectual curiosity'.)

Luckily they were roughly the same build, so they bought two identical sets of sweat shirts, pants and hats. They also bought two large plastic bags to wear underneath these clothes, so that any rain which ended up on their clothes would not get soaked into their underclothes. They then measured out a 100-metre track behind their United States National Climatic Data Center office and waited for some rain. Soon, some heavy rain came along — falling at around 18 mm (or 3/4 of an inch) per hour. They made sure that they weighed the clothes both *before* and *after* the rain.

Dr Wallis ran the hundred metres at around 4 m/s (about 14.4 kph, or 9 mph), and his clothes absorbed 130 grams (about 4.5 oz) of water. Dr Peterson walked his hundred metres at a much more leisurely 1.4 m/s (about 5 kph, or 3.1 mph), but his clothes soaked up 217 grams (about 7.7 oz) of water. *Running*, instead of *walking* meant that you got 40 per cent less wet, which was pretty darn close to their predicted 44 per cent.

Run! Run! Run!

So if you run in heavy rain (as compared with walking), you'll stay somewhere between 30 per cent and 50 per cent drier. The greatest benefit is achieved by running in heavy windy rainy conditions, and by leaning forward. There is less improvement in light rain, with no wind, and when you stay nearly vertical.

Of course, you could take an umbrella with you — but that would give you lousy aerodynamics. But it's probably safer than running on slippery ground. Then again . . . if you're riding a bicycle . . . if you're in love . . . if it's raining . . . you may just dare to get wet.

The Guinness Book of Records, CD-ROM, Guinness Publishing, 1993.

J.J. Holden, S.E. Belcher, A. Horvath and I. Pytharoulis, 'Raindrops Keep Falling on my Head', *Weather*, Vol. 50, November 1995, pp 367–370.

Thomas C. Peterson and Trevor W.R. Wallis, 'Running in the Rain', *Weather*, Vol. 52, March 1997, p 93.

Jearl Walker, *The Flying Circus of Physics*, John Wiley & Sons, 1975, pp 23, 231–232.

Martin Whittle, *New Scientist*, No. 1960, 14 January 1995, p 57.

Matthew Wright, *New Scientist*, No. 1960, 14 January 1995, p 57.

Mike Stevenson, 'Drip Dry', *New Scientist*, No. 1960, 14 January 1995, p 57.

NOAH'S FLOOD

Many societies have the myth of a 'Great Flood' that washed everything away. This myth exists in the folklore of the American Indians and of the Ancient Greeks, and in the Aztec, Jewish and Mesopotamian religions. This 'Great Flood' always happened a long time ago — in the early days of the human race. The myth of the 'Great Flood' also exists in the Bible, which tells the story of how Noah floated safely in his Ark above the rising floodwaters.

Stories about a great flood have circulated for many generations, and now some marine geologists think they have some science to back up the story of this 'Great Flood'. They claim that it was the sudden re-filling of the Black Sea.

God, Noah, and the Ark

The Book of Genesis in the Bible tells that the Earth's population had increased since Adam and Eve were expelled from the Garden of Eden. The society was relatively advanced — Tubal-Cain *'ancestor of all those who work copper and iron'* (Gen. 4:22) — had even started off the science of metallurgy. According to the Bible, however, the human race had lost sight of their God and had fallen into sin. So God said, *'I shall rid the surface of the Earth of the humans whom I created ... for I regret having made them'* (Gen. 6:7–8).

But not all people were bad. Noah was a righteous man who *'found favour in the eyes of the Lord'* (Gen. 6:8). So just before God was about to wash the bad people off the face of the planet, he gave Noah a personal warning, telling him to build an Ark. This Ark was huge, with a length of *'300 cubits, its breadth 50 cubits, and its height 30 cubits'* (Gen. 6:15). In our measurements, this works out to about 150 metres long, 25 metres wide, and 16 metres high. (According to Genesis, Noah was pretty cool — not only did he restock all the animals on the Earth today, but his sperm gave life to everybody alive today via his three sons, and he also invented the modern vineyard.)

Noah floated on the floodwaters for 40 days and 40 nights. He then sent out some birds, and when one finally returned with an olive branch, this showed that the land was exposed again, and so Noah and his family 'landed' and left the Ark to start civilisation going again.

The 'Black Sea Theory'

First of all, we can ignore the idea of a flood big enough to cover the whole planet to a level as high as Mount Ararat in Turkey — there's simply not enough water on Earth. But what about a *local* great flood?

A team of marine geologists have given their explanation of the Great Flood myth in the April 1997 issue of a magazine called *Marine Geology*. The team was led by Kazimieras Shimkus, from the Shirshov Oceanology Institute in Russia, but William Ryan and Walter C. Pitman III from the Lamont-Doherty Earth Observatory in New York, wrote the paper.

During an Ice Age, the ocean level is much lower than usual. According to these scientists, as the last Ice Age gradually wound down, and the ice melted, the ocean levels rose. They claim that the Great Flood could have been the Mediterranean Sea suddenly filling the Black Sea, about 7000 years ago.

Black Sea Geography

The Black Sea is a long way from the Atlantic Ocean. First, enter the Mediterranean Sea by passing the Rock of Gibraltar. Go past Italy, and right at the far end of the Mediterranean, you will come to the Aegean Sea. Between the Aegean Sea and the Black Sea, you will come to the water separating Greece and Turkey — and Europe from Asia.

There are actually three bodies of water between the Aegean Sea and the Black Sea, each with its own name. There is the skinny Dardanelles Strait, the much wider, but still small Sea of Marmara, and the tiny Strait of

THE GREEK FLOOD

In the Greek version of the Great Flood, Zeus decided to destroy the human race. But Deucalion built an Ark, and like Noah, he and his wife landed on a mountain.

They asked how to restart the human race, ignoring the obvious method! They were told to throw 'the bones of their mother' behind them. So, they threw behind them the stones of 'Mother Earth'. The stones of Deucalion turned into men, and the stones thrown by his wife, Pyrrha, turned into women.

Bosporus (about 30 kilometres long and one or two kilometres wide) which opens into the Black Sea.

The Black Sea has an area of about 422 000 square kilometres — roughly twice the size of Victoria, three times the size of England, and about the same size as California. At its deepest, it is about 2.2 kilometres deep.

The Black Sea is basically a freshwater sea that is contaminated with salty ocean water, so it is roughly half as salty as the oceans. In the Black Sea, very few fish live below about 150 metres, because the deeper waters are high in hydrogen sulphide and low in oxygen, and there is not much mixing between the shallow and deep waters. (However, there are special bacteria in the Black Sea, that have adapted to the hydrogen sulphide.)

The Theory

An article in *Marine Geology* by William Ryan and others, claims that the Black Sea partly dried out during the last Ice Age, and then suddenly re-filled causing major flooding.

The Background

The last Ice Age lasted for approximately 100 000 years, and wound down about 10 000 years ago. A lot of the land on the planet was covered with ice. During this Ice Age, the ice packs advanced until they covered the top of North America and much of Europe, and it was a few kilometres thick in Germany and New York. The water to make this ice came from the ocean, so the level of the ocean dropped by about 120 metres. Today, the Bosporus Strait is only about 35 metres deep. One hundred and twenty metres is a lot deeper than 35 metres, so during the last Ice Age, the Bosporus Strait would have been dry land, and the Black Sea would have been well and truly cut off from the Mediterranean Sea.

The Search for Evidence

The scientists think that shells of sea snails (both saltwater and freshwater) are the evidence that prove their claim about the Black Sea filling catastrophically.

In 1993, a group of Russian and American marine geologists took a converted fishing boat for a two-week exploration through the Black Sea. The Russians, Kazimieras Shimkus and Vladamir Moskalenko, wanted to examine the mud on the floor of the Black

Sea for radioactive fallout (from Chernobyl). The Americans, Pitman and Ryan, wanted to look for evidence of a sudden flood. So the scientists did a deal. The Americans helped the Russians look for radioactivity, and the Russians helped the Americans look for evidence of a flood.

One of their machines blasted sound-waves into the floor of the Black Sea. It then analysed the reflections, to make a map of the different layers of the silt and mud on the sea floor. The map of the different layers looked a little like a multi-layered cake. The map from the 'sound-blaster' told them where to use the second machine. This was a 'sucking' pipe, that they used to suck up material from the floor of

the Black Sea — including snail shells. In a few selected areas they examined the different layers of sediment on the floor of the Black Sea.

MEDITERRANEAN GREAT FLOODS

The Mediterranean Sea has come and gone as the continents have shifted. The continents (on their tectonic plates) move at roughly the rate that your fingernails grow — about five centimetres per year.

About six million years ago, Africa had moved far enough north to create an isthmus. This isthmus closed off the mouth of the Mediterranean Sea. Over the next thousand years, the water in the Mediterranean Sea (all four million cubic kilometres of it!) evaporated leaving behind dried salt. It stayed like this for about a million years.

(The climate around the Mediterranean is quite arid so about 4000 cubic kilometres of water evaporates each year from the Mediterranean. Only about 400 cubic kilometres of fresh water is added to the Mediterranean each year from rainfall and rivers. So today, the remaining 90 per cent of the evaporation is compensated for by seawater flowing in from the Atlantic Ocean.)

Then, about five million years ago, the Atlantic once again burst through the Strait at Gibraltar, and very rapidly turned the Mediterranean 'Basin' back into the Mediterranean 'Sea'. This was a magnificent event. The waterfall would have been enormous. The water would have rushed in at about 40 000 cubic kilometres per year — 100 times bigger than Victoria Falls, and 1000 times bigger than Niagara Falls. But at that time, there were no humans around to witness this particular Great Flood.

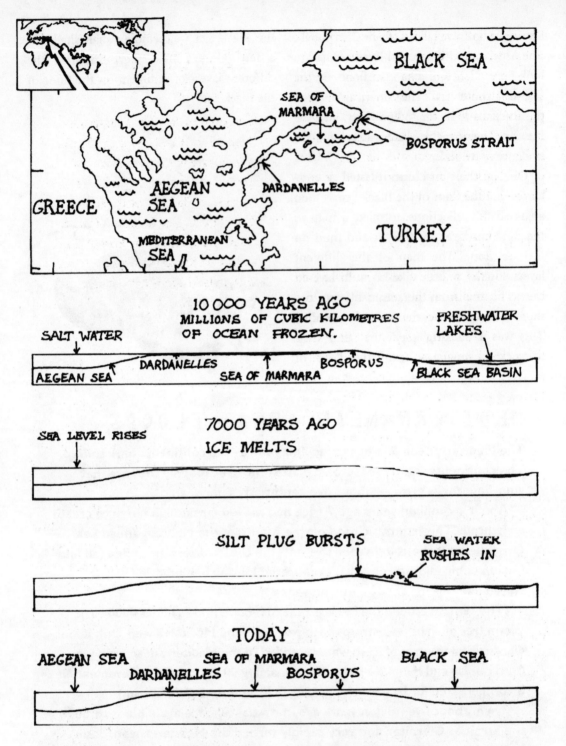

10 000 YEARS AGO
MILLIONS OF CUBIC KILOMETRES OF OCEAN FROZEN.

SALT WATER

FRESHWATER LAKES

AEGEAN SEA — DARDANELLES — SEA OF MARMARA — BOSPORUS — BLACK SEA BASIN

7000 YEARS AGO
ICE MELTS

SEA LEVEL RISES

SILT PLUG BURSTS

SEA WATER RUSHES IN

TODAY

AEGEAN SEA — DARDANELLES — SEA OF MARMARA — BOSPORUS — BLACK SEA

The Evidence and Proof

In the lowest level of the sediment, they found the type of dense mud that you usually find in river deltas. In the next level up, they found thick layers of freshwater snail shells. These shells were bashed and broken. This is exactly what you would expect to find in a shallow freshwater river, where the wind and waves batter the shells around.

Above the layer of broken freshwater snail shells, the scientists found a layer of unbroken saltwater snail shells. They were unbroken because these shells had always been *deep* beneath the saltwater waves. These saltwater snail shells had never been pushed around by shallow waters. The team measured the age of the saltwater snail shells at many different sites across the floor of the Black Sea and found that they were all about 7000 years old.

The Scenario

So this is the scientists' scenario based on the snail shells: during the Ice Age, the water level in the oceans was very low. The Bosporus Strait turned into just a little river that carried small amounts of water out of the Black Sea, and dumped it into the Mediterranean Sea. Soon this river became choked with silt and mud and ran dry. The Black Sea shrank. The deeper parts held water, but the shallow coastal margins were exposed. The Black Sea turned into a giant freshwater lake, with a few smaller lakes, which were fed by rivers. Freshwater snails lived in the new coastal shallows. When these freshwater snails died, their shells were broken by the wind and waves in the shallow fresh water. People moved in, and grew their vegetables in the fertile soil near these lakes and rivers. The area of this smaller Black Sea was about 100 000 square kilometres less than it is today, and the water level was about 100 metres lower than today.

At the end of the Ice Age, when the ice began to melt, the water levels rose in the Atlantic Ocean and in the Mediterranean Sea. They rose until they were level with the Bosporus, and kept on rising. About 7000

MIDDLE EAST FLOOD MYTH

A people called the Sumerians lived about 5500 years ago in that part of the world which is now Southern Iraq, which stretches from Baghdad to the Persian Gulf. These people have left us a document, called the Sumerian King List. It records eight kings, who ruled before the coming of the Great Flood.

years ago, the waters became so high that the water pressure blew out that plug of silt and mud that kept the Black Sea (or Black Sea Basin) separate from the Mediterranean. According to one of the authors of the paper, 'the waters would have come thundering in with the force of 200 Niagaras, and a roar that could be heard at least 100 kilometres away'. Ryan and Pitman calculate that the water flooded in at greater than 50 cubic kilometres per day. At first, the water level rose at some tens of centimetres each day. Within a few months, 100 000 square kilometres of fine agricultural land was underwater.

Snails - Key Evidence

The waters rose very rapidly. The freshwater snails died from the salt water and the saltwater snails moved in. The water was still fairly shallow, so the shells of the freshwater snails were soon broken by the wind and waves. As the water got deeper (and more salty), the saltwater snails took over. When the saltwater snails died, their shells survived better in the deeper water, because they were not battered around.

As the water flooded in, it forced the people living in the Black Sea Basin to move on. The soil around that freshwater sea had been very fertile, but the waters rose rapidly, so the inhabitants had to go. Some went north to Russia and the Ukraine, some went west to what is now Romania, Bulgaria and Greece, and some went east to Georgia and India. And some went south to Turkey and the Middle East.

And wherever they went, these people took the story of a Great Flood with them.

NAMES OF THE BLACK SEA

The ancient Greeks called the Black Sea 'Pontus Axeinus' — the 'Inhospitable Sea' — but as they became more familiar with it, and even set up colonies on its shores, they changed the name to 'Hospitable Sea' — 'Pontus Euxinus'. This was the sea that Jason crossed with his Argonauts, to search for the Golden Fleece in what is now Georgia.

When the Turks controlled this region, they hated the unpredictable storms that would suddenly whip up out of a calm sea, so they called it 'Karadeniz', the 'Black Sea'.

THE BABYLONIAN GREAT FLOOD

The Babylonian version of the Great Flood has many of the elements of the Biblical version – the man who is given a personal command from above, the torrential rains, the enormous ship, and the use of birds to test the land in the outside world.

The Babylonian epic story of Gilgamesh gives no explanation of why the gods had decided to destroy the human race. In fact, the only reason why the hero survives the flood is that one of the gods, who's a bit of a prankster, thinks it would be a good trick to play on the other gods.

The Gilgamesh epic was written on clay tablets way back around 700 BC, but the story comes from about 2000 BC. Some scholars think that both the Babylonian and Biblical stories originate from a single earlier story.

AUSTRALIAN GREAT FLOOD

Aborigines in Arnhem Land (at the top end of the Northern Territory) have their story of Wititj, the Olive Python. Wititj was at his waterhole, when the Wagilag Sisters disturbed him. Wititj started spitting in anger, and caused the first monsoon, which caused the first flood.

Certainly, the Aborigines have been in Australia for at least 40 000 years, and probably much longer. They would definitely have been here at the end of the last Ice Age, when the ice sheets began melting, and the ocean levels began rising. The rising ocean would have flooded about 20 per cent of Northern Australia.

Basin or Bladder?

Was there really, at some recent stage in human history, a Great Flood in the Black Sea so magnificent that it inspired flood myths for the next 7000 years? Pitman admits that, with specific written records, it's really hard to prove a link between Noah's Great Flood, and the great flooding of the Black Sea. Even so, he likes the theory.

Should we believe Pitman or should we believe the Hungarian psychological folklorist, Geza Roheim? He reckoned that the reason that there were so many myths about the Great Flood, was very simple — we humans simply dreamed about water, as our bladders gradually filled through the night!

REFERENCES

Kenneth J. Hsü, 'When the Mediterranean Dried Up', *Scientific American*, December 1972, pp 26–36.
—— 'When the Black Sea was Drained', *Scientific American*, May 1978, pp 53–63.
Rosie Mestel, 'Noah's Flood', *New Scientist*, No. 2102, 4 October 1997, pp 24–27.
William B.F. Ryan et al, 'An Abrupt Drowning of the Black Sea Shelf', *Marine Geology*, Vol. 138, 1997, pp 119–126.

CAN'T GET TO THE STARS IN A LIFETIME

In a relatively short time, we humans have made it from the Stone Age to the Space Age. However, our space travel technology is still quite primitive. We can send our robot spacecraft beyond the planets of the solar system, but humans can barely get to the moon.

Some people say that it will always be impossible to get to the nearest stars and back in a human lifetime. But maybe our lifetimes will get longer, or our technology will get better.

However, with our current technology, we're hard-pushed to get anywhere near the stars. The very best we could do is to send a four-gram payload on a one-way trip — and it would take 20 years to get to the nearest stars. And to get that tiny four-gram payload to the stars, we would need to construct, and run, the biggest laser ever built. The laser would be so big that it would take one per cent of the electrical power that the entire human race could generate by the end of the 20th century!

Einstein Says No

If you want to travel around the universe rapidly, you come smack up against Einstein. Albert Einstein threw a few spanners in the works with his Special Theory of Relativity.

First, he said that nothing can travel faster than the speed of light. To be more accurate, you can't send either matter or information faster than the speed of light, which is around 300 000 kilometres in one second (about eight times around the Earth in one second).

Second, Einstein theorised that strange things happen as you travel at close to the speed of light — and he was right!

As you speed up, you get heavier! So you need more energy to make your extra mass go faster. But the faster you go, the more energy is used up, making you heavier, not faster. By the time you get up to the speed of light, you would have an infinite mass, which is ridiculous! (The only particles that travel at the speed of light, such as the photon, don't have any mass at all.)

The other strange thing is that as you go faster, your time slows down and you literally get shorter. In *Star Trek*, the starship *Enterprise* can travel at around one-quarter of the speed of light. At that speed, it is about 3 per cent shorter, and its clocks go about 3 per cent slower than our clocks back here on Earth.

It's a Long Way to the Stars

With our present-day science, it seems awfully hard to get to even the nearest stars in a reasonable time. There are three stars in the Alpha Centauri complex, which is about 4.3 light-years away. A light-year is the distance that light travels in one year, and it's a really long way. For example, the planet Neptune is only about four light-*hours* from the Sun, and it took the *Voyager* spacecraft 12 *years* to get there.

So how long would it take to get to Alpha Centauri?

At 100 kilometres per hour, we can get to the Centauri stars in 50 million years. At the speed that the *Apollo* astronauts used to go to the Moon, that time is cut down to only 900 000 years. And if we travel at the speed of the *Voyager II* spacecraft, our journey time is 80 000 years.

The Egyptian, Jewish and Chinese civilisations all got started about 5000 years ago. Eighty thousand years is about 160 times longer than that. Who would still be alive back home to send your postcards to?

But suppose that we could make a starship travel 80 times *faster* than the *Voyager II* spacecraft. Suppose that we could invent a hibernation process for human space travellers that could keep them in 'cold storage' for 1000 years. Chocolate and other essential foods would also be stored for 1000 years, so the starship travellers would have something to chase away those cold-storage blues when they got to the stars.

Suppose that we wanted to send the astronauts in a vehicle roughly the size of the American space shuttle. Of course, the starship would need fuel. Using conventional rocket fuels, the amount needed would be much greater than all of the mass of the universe — in fact, over 10 000 million million million million million million million million million million times greater than all the mass in the entire known universe!

So even if we want to send a starship the size of the space shuttle to the nearest stars, in a journey taking 1000 years — there would not be enough matter in the universe to power our starship with conventional rocket fuels.

Far-out Ways to the Stars

There are other ideas about how we might get to the stars, based on the science that does exist today — but they are pretty way-out!

Back in the 1950s, scientists came up with Project Orion, a strange starship which basically involves strapping nuclear bombs to your backside! At the back end, the starship had an enormous 'pusher plate', mounted on huge shock-absorber springs. The plan was to explode five nuclear weapons every second, which would push on the giant plate — and propel the starship forward, in a series of nuclear 'putt-putts'.

In 1973, the British Interplanetary Society came up with Project Daedalus (see picture on page 34). It was similar to Project Orion, but instead of five nuclear explosions every second, they wanted to have 250 micro-fusion nuclear explosions every second. They reckoned that they would be able to accelerate a pay-load of 100 tonnes up to about 12 per cent of the speed of light. So you could get to the Centauri complex in about half a century. The big engineering problem then and now is that we can't make tiny fusion explosions yet. Another challenge was that the starship had to pick up its fuel, Helium–3, from the atmosphere of Jupiter, on the way out of the solar system.

Bussard came up with another way when he invented the Interstellar Ramjet in 1960. The Interstellar Ramjet would not carry its own fuel — it would scoop up fuel from the space that it was travelling through. It had a funnel out the front, about 1000 kilometres across. This funnel was not solid, but made of electromagnetic fields that were generated by an enormous magnetic solenoid (see picture on page 34). The plan was to trap particles in the funnel, feed them to the fusion reactor (not invented yet), burn them and throw them out of the back of the Bussard Ramjet.

Robert Forward came up with Interstellar Laser Sails in the 1980s. His idea was to leave your power supply at home, by using '*light, rather than rockets*'. After all, light can exert a very tiny push on an object. If you have a very powerful light shining on a very large object, the amount of 'push' gets reasonably big.

His first plan proposed shining a 10 million gigawatt laser on a sail 1000 kilometres across. According to his calculations, he could send a vehicle weighing 1000 tonnes to the Centauri complex in just 10 years. Catch number one was the 10 million gigawatt laser — 10 million gigawatts is 10 000 times more than all of the power generated on the Earth today!

Catch number two was that if you were one light-year out from Earth, and you needed to slightly correct your course, it would take two years to change your course — one year for the message to get back to the laser, and another year for the changed laser light to reach your Interstellar Laser Sails.

So Robert Forward came up with a smaller version. The new sail had a grid of fine wires weighing only 16 grams, but covering a kilometre. This time, the laser was only 10 gigawatts, which is only one-hundredth of all the power generated on the Earth today. The penalty is that the trip would take twice as long (20 years), and the payload would not be 1000 tonnes, but only 4 grams!

And of course, for either version it would be a one-way trip.

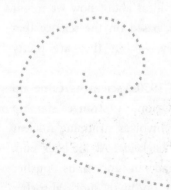

HOW TO MEASURE SPEED OF LIGHT

Light travels awfully fast — around 300 000 kilometres in one second.

Galileo was one of the first scientists to try to measure the speed of light. He sent a colleague with a lantern to a nearby hill. As soon as Galileo saw a flash of light come from his friend's lantern, he immediately flashed his lantern back again.

All they did was measure each other's reaction times, and prove that light was really really really fast!

In the 1670s, a Danish scientist called Ole Roemer stumbled across another way to measure the speed of light. He was looking at the moons of Jupiter as they went behind the planet. He saw that these eclipses were happening later, as the Earth was travelling further away from Jupiter, and that the eclipses happened earlier, when the Earth was travelling closer to Jupiter.

Earth is fairly close to the Sun, while Jupiter is about five times further out. When the Earth and Jupiter were on the same side of the Sun, the eclipses happened sooner. When they were on opposite sides of the Sun, the eclipses happened later.

Roemer worked out that the light from Jupiter had to travel a greater distance to get to the Earth when it was on the far side of the Sun. His measurements showed that light took around 1000 seconds to cross a circle the size of the diameter of the Earth's orbit around the Sun.

In 1923, Albert A. Michelson got the United States Coast and Geodetic Survey to measure very accurately a base line in the San Gabriel Valley, in Southern California. At that time, it was the most accurately measured line that had ever been laid out on the surface of the Earth. He set up a rotating mirror on Mount Wilson, and a fixed mirror on another mountain 22 miles away. He came up with a measurement for the speed of light that was within fractions of a per cent of what we have today.

New Ideas are Hard

So you can see that we are not going to get to the stars and back in a human lifetime with our present technology. We need to make a big jump.

We didn't invent photocopiers by trying to make better carbon paper. We didn't invent transistors by making better vacuum valves. We didn't invent steamships by improving sails and rigging.

Instead, we made breakthroughs in our technology. We went from the sailing ship, to the steamship, to the propeller plane, then the jet plane, and today we have the rocket. We need to make the breakthrough from the rocket to the next stage.

This is just a personal opinion, but I think it is possible to make that breakthrough that will get us to the stars and back in a human lifetime. Some NASA scientists think so too.

CAN GET TO THE STARS IN A LIFETIME

All the other stories in this book are solidly based on proven science. This story is different — it's guessing about the future. Only time will tell if my guesses are even slightly correct.

Every time I watch the show *Star Trek*, I wish that I could travel in their wonderful starship, the *Enterprise*.

The *Enterprise* has two types of travel. Impulse Drive is fairly slow (less than the speed of light), and is used when you're close to your final destination. The *Enterprise* can accelerate almost instantly from zero to Impulse Speed, which is about one-half of the speed of light. Warp Drive is very fast (faster than the speed of light) and is used for covering large distances rapidly. At flat chat in Warp Drive, the *Enterprise* can travel at 2000 times the speed of light. You could certainly get a close look at many of the nearby stars if you could travel at such enormous speeds. It would get you from one side of our galaxy, the Milky Way, to the other side in just 50 years.

Unfortunately, the *Enterprise* exists only on *Star Trek*. We have to get to the stars by ourselves.

Can we ever get to the stars in a human lifetime? The answer depends on whom you ask:

No — according to the 'Old Physics'.

Possibly — according to a NASA team, which has identified three breakthroughs that could get us to the stars.

Probably — according to the 'New Physics' (which has not been invented yet).

If we're going to get to the stars in a human lifetime, we can't get there using conventional engineering and physics. We're going to need some radically new physics to get to the stars. Some, but definitely not all, physicists think that a new physics is just around the corner.

Problems in Physics

The 'feel' in Physics today is a bit like the situation that existed a century ago. Around 1900, there were a few problems in the Land Of Science. These problems were too hard to solve, and so the physicists just ignored them — they swept them under the carpet.

One of these problems was the Photoelectric Effect. It was known that if you shone light onto some special materials, they would give off electricity. But the strange thing was that this Photoelectric Effect would work with blue light, but not red light! It took Albert

Einstein to work out why, and he won his Nobel Prize for this work. His answer lay in the New Physics, not the Old Physics of the time.

Another problem was the age of the Sun. The *geologists* said that the Earth had to be at least 25 million years old, and was probably a lot more. The *astronomers* knew how big the Sun was. But the best fuel that the *physicists* could come up with was coal. If the Sun was made entirely from coal, it was big enough to burn for only a million years. How could the Sun have kept on burning for the extra 24 million years?

Scientists came up with bizarre theories like the Sun getting extra heat energy from millions of comets continually ramming into it at high speed — but basically, they swept the problem under the Cosmic Carpet. Early in the 20th century, nuclear energy was discovered, and the problem was solved. But the problem had to be solved with a New Physics, not the Old Physics.

Today, nearing the end of the 20th century, some physicists think that we're in a similar situation. There are quite a few problems that we sweep under the carpet and ignore, simply because we can't solve them. One of these little problems is the missing mass of the universe — according to the astronomers, 95 per cent of the universe is 'missing'! This means that we can measure its effects so we know that it is there, but we can't see it with any of our telescopes.

Today, many scientists think that because of these types of big questions, and many other problems, we're heading for another revolution in Physics.

NASA's Plan

NASA thinks so too. Over the last few years NASA has quietly organised some extraordinary science projects. One of them has the seemingly harmless title of 'Breakthrough Propulsion Physics Program'. What it's really about, is coming up with ways (both short term and medium term), of making the breakthroughs that will give us travel to the stars and back in a normal human lifespan.

The Three Breakthroughs

In 1996, Marc Millis of the NASA Lewis Research Center created the Breakthrough Propulsion Physics Program. This program has a small floating group of scientists from government, university and industry. Since 1996, they've banged their heads together, and worked out what needs to be done.

A NASA team thinks that they have identified the three breakthroughs needed. The breakthroughs are: to get rid of the need for propellant; to be able to travel much faster; and to harness new sources of energy.

1. Less or No Propellant/Warp Drive

The first breakthrough the scientists are thinking about is getting rid of, or at least dramatically reducing, the need for propellant, or fuel.

The three main approaches to this breakthrough are to carry less propellant, to carry no propellant, or to bypass the need for any propellant at all.

① REDUCE PROPELLANT.

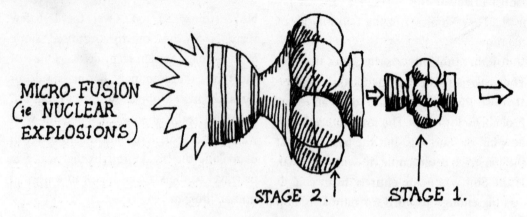

MICRO-FUSION
(ie NUCLEAR
EXPLOSIONS)

STAGE 2.   STAGE 1.

② DON'T CARRY PROPELLANT.

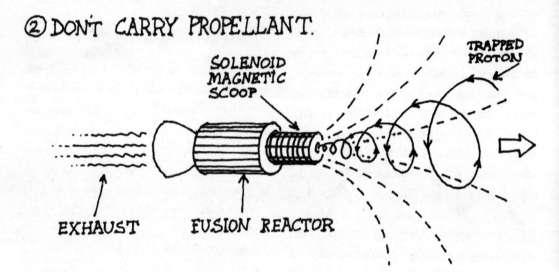

SOLENOID
MAGNETIC
SCOOP

TRAPPED
PROTON

EXHAUST   FUSION REACTOR

③ FORGET PROPELLANT, WARP SPACE INSTEAD.

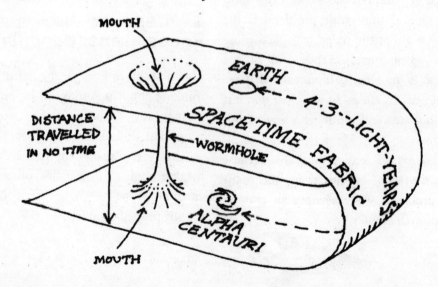

MOUTH

EARTH

---4·3·LIGHT-YEARS

SPACETIME FABRIC

DISTANCE
TRAVELLED
IN NO TIME

WORMHOLE

ALPHA
CENTAURI

MOUTH

Rockets which propel spacecraft work by Newton's Third Law of Motion — for every action, there is an equal and opposite reaction. When you drive your car down the road, your wheels grip the road, and your engine turns the wheels. But in space, there's nothing to grip onto. That's why rockets (which go forward by tossing propellant out the back) are the only practical way we have of travelling through space — at least, at the moment.

But then you get caught in a vicious cycle — propellant has mass, and you need extra propellant to accelerate your first lot of propellant. Things can quickly get out of hand. For example, to get the Space Shuttle to the nearby Centauri stars, in a thousand-year journey and using conventional chemical fuels, you need to throw away huge amounts of propellant — squillions (10^{64}) of times greater than the mass of the entire universe!

1.1 Reduce Propellant

So some scientists are trying to *reduce the amount of propellant* needed, by thinking about more powerful propellants, such as nuclear fission, nuclear fusion, or even anti-matter. (Matter has a positive core with negative electrons running around it and anti-matter is the opposite — a negative core with positive electrons.)

To get to the Centauri stars in 1000 years with a nuclear *fission* rocket, you would need about a billion super-tankers of propellant, each weighing about a quarter of a million tonnes. If you use a nuclear *fusion* rocket, you'd need only about 1000 super-tankers. But with an *anti-matter* rocket, you'd need only about 10 railway tankers. But over the last half-century, we have made only one-tenth of a billionth of a gram of anti-matter — which is barely enough to get

your fully-loaded car down the driveway before it splutters to a high-tech halt!

1.2 Don't Carry Propellant

Another way to reduce the propellant carried is *not to carry it* at all!

You could pick up your propellant as you travel (via a giant funnel). This is the idea behind Bussard's Ramjet.

Another way to not carry propellant is to have a 'pushing' machine back in the Solar System. Bob Forward plans to use an enormous laser to 'push' a spacecraft away.

1.3 Forget Propellant, Warp Space Instead

But other scientists want to get rid of the propellant entirely. They are thinking about how to manipulate gravity or inertia, or how to interact directly between matter and the spacetime fabric of the universe. Here, they may be able to travel faster than the speed of light (let's forget Einstein for a moment)!

Miguel Alcubierre, a physicist at the University of Wales in Cardiff, worked out (theoretically) how to 'warp' spacetime, and travel faster than the speed of light. All you have to do is shrink the fabric of spacetime in front of your spaceship, and at the same time expand it behind your spaceship. (Scientists use the word 'warp' to mean 'bend', 'shrink' or 'expand'.)

Let me give you an example. Suppose that you want to leave Earth, and go to the nearest star system, Alpha Centauri. Suppose that you are already in a low Earth orbit — about 300 kilometres up. So you are 300 kilometres from Earth, and 4.3 light-years from Alpha Centauri. Suppose that you had a magic machine that could shrink, or expand, any bit of spacetime that you wanted.

All you have to do is expand the space between you and the Earth (from 300 kilometres to 4.3 light-years), and shrink the space between you and Alpha

THE FABRIC OF SPACETIME

The 'fabric of spacetime' sounds complicated, but it isn't really. It's just the usual three dimensions of space, with the added fourth dimension of time.

Look at a sheet of paper. It has two dimensions — left–right and up–down.

The 'space' part of our real world has three dimensions — left–right, up–down and backwards–forwards. So three dimensions are enough to tell you where on the planet your house is.

But if you want to meet somebody at the movies you need a fourth dimension — time. (Time normally ticks away at the rate of one second per second.)

Spacetime is just these four dimensions — left–right, up–down, backwards–forwards and time. The fabric of spacetime defines where and when we are in our real world.

Centauri (from 4.3 light-years to a safe distance from Alpha Centauri) — and suddenly you would have left the Earth, and arrived at Alpha Centauri. If your journey took less than 4.3 years, you would have travelled faster than the speed of light.

Let's demonstrate this in the safety of your own home! Take a sheet of paper. Let's pretend that the fabric of spacetime is a sheet of paper, and that our starship is a pencil. Einstein tells us that our pencil (the starship) can't move faster than the speed of light. But Einstein doesn't mind one bit that our sheet of paper (the spacetime fabric) can itself move faster than the speed of light!

After all, back in the good old days, just after the Big Bang happened, it's claimed that the fabric of spacetime itself expanded faster than the speed of light. All Einstein says is that nothing can move faster than the speed of light, if it exists by having the fabric of spacetime as part of its background. But the *fabric* of spacetime itself can move at any speed.

It seems that Miguel Alcubierre is correct as far as he went, but he didn't go far enough.

Mitchell Pfenning and Larry Ford from Tufts University in Medford in Massachusetts followed on from his work, and actually figured out a few more numbers. They think that he is correct, but there are a few problems. First, they found that you could warp only a very tiny section of spacetime (much smaller than an atom). Unfortunately, your spaceship will have to be bigger than a single atom. Second, they found that the energy needed to warp this tiny bit of spacetime is around 10 billion

times all the energy in the entire Universe — hardly worth it!

Another way to get moving faster than the speed of light is to use a 'wormhole'. A wormhole joins one place to another. According to the physicists, a wormhole is a perfectly reasonable mathematical solution to some problems in physics (but of course, we haven't built one yet, or seen one, or proved that they exist).

Here's how you could use a wormhole to get from one place to another, without crossing the distance in between.

Let's imagine that there are two-dimensional people who live on a sheet of paper, called Flatland. These people know of only two dimensions — 'north–south' and 'east–west'. They can't move in the third dimension (which would take them out of the paper). Suppose that there is a little black dot at the top right-hand corner of the page, and another little black dot in the bottom left-hand corner. When the Flatlanders want to get from one dot to the other, they have to travel all the way from one corner of the paper to the other, and they have to cross all the distance in between.

But suppose that a clever Flatlander invents two machines. One of them 'bends' or warps their two-dimensional space. The other machine takes them into the mysterious third dimension. Together these two machines can make a 'wormhole'.

Our clever Flatlander uses the first machine. It 'bends' the sheet of paper (the Flatlander spacetime), and brings the two dots in contact with each other. The second machine then manipulates the third dimension. Our Flatlander can jump out of the paper at the bottom dot, and back into the paper again at the top dot. Our Flatlander has just travelled from one dot to the other, and didn't cross the space in between.

A wormhole could do the same thing, but in our four-dimensional spacetime universe.

Notice that with both the 'warping of space' method, and the 'wormhole' method, we're no longer throwing propellant out of the back of our spaceship.

HOW TO THINK ABOUT OUR FUTURE

There is a technology spectrum. It runs from conjecture to speculation to science, then technology and finally application.

Conjecture is where we begin our search for knowledge. We know that we would like to get to the stars, but we don't know if it's even possible.

At the moment we're in the speculation stage. We know what we do know, and what we don't know, and have some ideas of where to go looking to solve the problems.

Still ahead of us is the stage of science where we develop our theories. At this stage, we will know whether it can be done, and what it will involve.

The next stage is technology. Here we actually begin to build the machines to get us to the stars.

Application is the final stage, when the technology is common enough to be used by much of the population. Telephones and jets are at this stage.

2. Go Very Quickly

The second breakthrough needed is to travel more quickly. In the days of sailing ships, the sailors could put up with a few years away from home. We should aim for getting to the stars and back in a human lifetime, but five years would be a lot better.

3. Lots of Grunt

The third breakthrough needed for real space travel is a lot more energy. At the moment, we use chemical energy, but nuclear fission, nuclear fusion or even anti-matter would give us much more grunt. Anti-matter is the most powerful of all of these forms of energy.

But, even so, using anti-matter to accelerate a starship from zero to half the speed of light, and back to zero, would take 700 times the mass of the spaceship. (That's a lot of anti-matter to carry around.)

There has to be a better way. We need some fundamentally new ways of generating energy.

There are other energy sources on the horizon.

There's a strange energy called Zero Point Energy. This is the fundamental energy of a vacuum. There's enough Zero Point Energy in one cubic centimetre of vacuum to boil all of the oceans on our planet! This is the level of energy we need to get to the stars and back in a lifetime.

But there are other energy sources that we haven't begun to tame.

The Four Forces

As far as the physicists are concerned, the natural universe can be explained in terms of the Four Forces. These forces are the Electromagnetic Force (radio, TV, telephones, your hi-fi unit, and the like), the Gravity Force (which keeps the planets spinning in their orbits around the Sun, and keeps you stuck to the ground), the Weak Nuclear Force (which is responsible for some forms of radioactivity) and the Strong Nuclear Force (which holds the positively charged particles in the core of the atom together).

At the end of the 20th century, we humans can manipulate only one of these four forces — the Electromagnetic Force. We can measure the waves and the particles that make up the Electromagnetic Force, and we can generate and block the Electromagnetic Force. Today we can use it to talk on a mobile phone to somebody on the other side of the planet, almost instantly.

Only 400 years ago we could hardly use the Electromagnetic Force at all. The most sophisticated thing that we could do with the Electromagnetic Force was to cut an orange in half, stick a copper nail into one side of the cut orange and an iron nail into the other side of the cut orange, and touch the two different metals to the wet leg of a frog. The leg of the frog would then jump as the electricity went through it. We have made more than a giant leapfrog in the last 400 years, in understanding and using the Electromagnetic Force.

Strange Unused Forces

Our ability to fool around with the remaining three forces of the universe is not even at that primitive stage, where we were with the Electromagnetic Force 400 years ago. We have a long way to go in our knowledge and our understanding of the other three forces — and somewhere along that pathway to understanding, we may uncover the secrets which will get us to the stars.

Weak Nuclear Force

You might hear people talk of the Unified Field Theory. The Unified Field Theory is the theory which links all of the four forces. We haven't so far managed to link all the four forces. We have, however, theoretically linked two of them, the weak nuclear force and the electromagnetic force. We now know that they are each different parts of the combined electroweak force.

As yet, at the end of the 20th century, we can't build machines using this knowledge. We don't know how to use the electromagnetic force to fool around with the weak nuclear force. If we did, we'd be able to control nuclear reactions with a skill that we don't have today.

Suppose you have a few grams of uranium. Every second, some of the atoms in the uranium will decay. Today there is nothing that we can do to speed up, or slow down, that process. But the fact that there is a link between the electromagnetic and the weak nuclear forces means that it might be possible to switch off a nuclear weapon before it explodes.

Take another example. Today we have nuclear power plants. Yet they generate nuclear power in a very primitive way. We put a few pieces of uranium close together, and they get very hot. We pour cold water over the uranium, and it turns into steam. We use that steam to turn the blades of a turbine, which makes electricity.

All we are using is the heat from the uranium. We could just as easily have got our heat by burning banknotes, furniture or paintings. We do not truly understand the weak nuclear force. If we did, we could get energy from the uranium directly, without having to go through all those complicated steps.

Imagine just getting 3 grams of uranium (about one-tenth of an ounce) in a little lead box. You take it home, and plug it into the electrical circuits of your house, via another small box — the Kruszelnicki Weak Force Transducer (imagined, but not yet invented). It gives you all the electricity

you need, for about 20 years. At the end of that time, it has turned to lead — which you take back to the shop, and exchange for another 3 grams of uranium. Understanding and using the Weak Nuclear Force could give us this capability.

Who knows what we could do once we truly understand the Strong Nuclear Force. Meanwhile there is another more familiar force waiting for us to understand.

Gravity Force

At the moment we don't understand gravity. There is no real way that we can control or manipulate gravity.

When we fly in a jet, we stay aloft because of the lesser air pressure on the top of the wing. The wing is held together by the Electromagnetic Forces. The air molecules are held together by Electromagnetic Forces. So in jets, we don't use or control the Gravity Force at all.

About the only case where we use the Gravity Force is in a pendulum. As it swings back-and-forth, it converts gravitational energy (at the top of the swing) to kinetic energy (at the bottom of the swing). We also use gravity in a primitive way in a hydro-electric plant.

A rocket does not control the Gravity Force. A rocket is a few metal tubes loaded up with a few hundred (or thousand) tonnes of explosive. Instead of letting the explosive burn in a few thousandths of a second, we stretch out the burning over 8 or 10 minutes. The burning of these chemicals is controlled by the Electromagnetic Force, which holds the atoms in the chemicals together. A rocket is a very dangerous machine, and it's a testimonial to the cleverness of the people who design and build them that only a few rockets explode.

Once the rocket gets into orbit, it has a huge amount of gravitational energy. That amount of gravitational energy is exactly the amount needed to put the rocket into orbit! We should put this energy into a Kruszelnicki Gravity Battery (imagined, but not invented yet) and use it to lift our next rocket into orbit. But at the moment, we waste that energy and burn it up as heat.

I believe that one day, we will control the Gravity Force. Imagine that you need to shift a 90-kilogram refrigerator. You just get the Kruszelnicki Gravity Neutraliser (not yet invented), place it on top of the fridge, and dial up 85 kilograms. Suddenly, the fridge weighs only 5 kilograms. You can now carry the fridge easily down to your Rent-A-Truck, so you can shift it to your new house.

Will We Do It?

Some breakthroughs happen rapidly. There were only three decades from 1911 (when we first began to understand radioactive decay) to 1942 (when the first working nuclear reactor was built under a gymnasium at Chicago University). It was only in 1997 that the Zero Point Energy was actually measured for the first time. Maybe it will take only three more decades before we begin to manipulate Zero Point Energy.

Why should we try to get to the stars? Well, one answer is that it's incredibly cheap, as compared to the annual military budget for the world. Another answer is that we are investing in the future for our children, and for the human race and our planet. The stars are our destiny.

REFERENCES

Marcus Chown, 'How to Pull a Fast One', *New Scientist*, No. 2098, 6 September 1997, p 49.

Gerald Feinberg, 'Particles That Go Faster Than Light', *Scientific American*, February 1970, pp 69–77.

Robert Matthews, 'Warp Factor Zero', *New Scientist*, No. 2092, 26 July 1997, p 6.

J.H. Rush, 'The Speed of Light', *Scientific American*, August 1955, pp 62–66.

http://www.lerc.nasa.gov/WWW/PAO/warp.htm

http://www.lerc.nasa.gov/WWW/bpp/

COFFEE, CAFFEINE AND CIRCLES

Coffee carries one of the world's most popular legal drugs — caffeine. Coffee has been around for at least a thousand years. We have learnt that a few cups of coffee per day won't hurt you, and it may even be good for you. But it's only in the last year that we have finally solved the mystery of why, when you spill coffee and it dries out, you get a dark ring on the perimeter of the spill, instead of a nice even colour all across the spill.

Popular Coffee

One-third of all the humans on the planet drink coffee. It's the second most popular drink after tea. In the USA, each person drinks about 100 litres (27 US gallons) per year. Coffee is attractive because it smells really nice, tastes OK, and stimulates you.

The smell and taste can be described using words such as aroma, body, flavour, taste, and aftertaste (just like the words used to describe wine).

Caffeine

But why coffee is so stimulating can be summed up in *one* word — caffeine.

Each cup of 'real' coffee (that is, coffee made from ground, roasted coffee beans rather than 'instant' coffee) can have anything from 80 to 350 mg of caffeine — depending on how much ground coffee you start with, how finely ground it is and the strain of bean used (*Arabica* or *Robusta*). On average, a cup of real coffee has about 100 mg, while a cup of instant coffee has a bit less (about 80 mg). However, espresso and Turkish coffee can have twice as much caffeine as this.

There is slightly more caffeine in coffee that has been made in a drip filter (115 mg), than in a percolator (85 mg). One theory about why this is, is that some of the caffeine is re-absorbed from the liquid coffee, back into the grounds on each pass. Another theory is that the coffee used in a drip filter is ground more finely than coffee used in a percolator — and the smaller particles of coffee in drip-filter coffee let the grounds give up more of their caffeine.

Decaffeinated coffee has only 3 mg of caffeine. A cup of tea has about 50 mg, which is about the same as a can of a fizzy cola drink. A 200 gram chocolate bar has about 40 mg of caffeine, while a drink of cocoa or hot chocolate has between 10 and 70 mg.

On average, around the whole world, each person takes about 70 mg of caffeine per day — about 54 per cent in coffee, 43 per cent in tea, and 3 per cent in other forms.

Caffeine Good

The drug effects of caffeine are really quite good for most people. It will boost your endurance by delaying the onset of fatigue — so you can run or cycle a lot further. It improves your memory, because it keeps your brain constantly alert and paying attention, so you can store information better in your long-term memory storage. It also helps you breathe a tiny bit more easily, because it slightly opens up your airways.

Caffeine will also reduce your chances of kidney stones, and cancer of the colon and rectum. And of course, that morning cup of coffee will kick your bowels into action, and keep you nice and regular (see the box on page 56 for more details).

Caffeine Bad

Caffeine did not deserve the harsh comments of Dr Kellogg (Mr Cornflakes) when he said in 1897, 'the nervousness and peevishness of our times are chiefly attributable to tea and coffee'.

Caffeine can interfere with your sleep (definitely true), and can sometimes increase your risk of heart disease (may be true). It can upset a stomach ulcer, and can sometimes cause gastric reflux (where the contents of your stomach come back up your oesophagus, and you can taste the acid in your mouth).

There are a few more known risks from coffee. First, coffee has fats or oils in it that raise your blood cholesterol level. According to Professor Martijn Katan from Wageningen University in the Netherlands, these oils are 60 times more powerful than butter, in raising your blood cholesterol levels! But if you make drip-filter coffee, so that the coffee liquid has to run through paper, the paper absorbs these nasty oils. (Actually, some scientists think that the real culprit is an alcohol, called cafestrol, that is present in the oil.)

Pregnant women are at special risk from a few cups of coffee per day. Having more than 300 mg of caffeine doubles the risk of having a baby under 2.5 kg (5.5 lb), and also increases the risk of a miscarriage.

Finally, a few people seem to be very sensitive to coffee, and find that caffeine makes them feel very anxious (see the box on page 57 for more details).

In summary, caffeine is amazingly safe if you keep your consumption under about five cups per day.

MOCHA

Mocha coffee gets its name from the port of Mokha (also called Mukha or Al-Mukha) not from chocolate as some believe. This port is on the Red Sea in the south-western part of Yemen. For many years it was the chief coffee-exporting port of all of Arabia — so the word 'mocha' came into the English language, to mean a high-quality coffee.

THE COFFEE TREE

The coffee tree belongs to the family *Rubiaceae* (madder family), *genus coffea*. This small tree is an evergreen. There are about 10 billion coffee trees under cultivation. Most of them are *Arabica* or *Robusta* strains.

When a tree is about five years old, it begins to bear fruit. The process starts with small white blossoms. When the tree is in full flower, it looks as though it is dusted with a light snow. About seven (*Arabica*) or nine (*Robusta*) months after flowering, the red berries ripen. There are two green coffee beans inside the red berries.

When a tree is around eight years old, its beans are now suitable for commercial markets. For the next 20 years, the coffee tree will produce commercial beans, and then the crop begins to reduce. An average coffee tree will give about 2 kg (5 pounds) of red berries which yield about 500 grams (1 pound) of green coffee beans per year.

The coffee tree grows best between the Tropic of Cancer and the Tropic of Capricorn and prefers a warm climate (23–28 degrees C, or 73–82 degrees F), and quite a bit of rain (1.8 metres or 70 inches). It will not survive snow, or a drought. The timing of the rainfall is important. It is better to have a heavier rain early in the season while the fruit is developing, but a lighter rain later when the fruit is ripening. In the wild, a coffee tree will grow up to 10 metres (33 feet) tall. But in coffee plantations, it is normally no higher than 6 metres (20 feet).

ARABICA VS ROBUSTA

There are about 20 to 29 different strains of coffee trees. But about 99 per cent of our coffee comes from just two strains — the *Arabica* strain and the *Robusta* strain. Hardly any coffee is made from a single crop — it's usually blended from several different lots.

ARABICA

The *Arabica* strain came from Ethiopia (and maybe Kaldi, the Arab goat herder — you'll hear about him soon). It has about 0.8 per cent to 1.5 per cent caffeine by weight. It's the main coffee grown in Central and South America, but is also grown in Indonesia, Asia and India. It thrives at high altitudes above 600 metres (2000 feet), where there are fewer invaders (insects, bacteria, fungi).

Arabica falls to the ground, and quickly goes bad. So it must be picked from the tree when it's ripe, before it has had a chance to fall and spoil. This means that the pickers have to visit the tree many times to pick the berries — which adds to the cost of the coffee.

Most experts say that *Arabica* has more flavour and aroma than *Robusta*. *Arabica* coffee is milder than *Robusta*. It is also more expensive, but dedicated coffee drinkers are happy to pay extra for the supposedly superior flavour of mountain-grown coffee.

ROBUSTA

The *Robusta* strain originally came from the Congo River Basin and East Africa. *Robusta* has less flavour than *Arabica*, but more kick. It has between 1.6 and 2.5 per cent caffeine by weight. It will survive warm humid climates better than *Arabica*. It's the main coffee grown in Africa, but is also grown in Indonesia, Asia and India. *Robusta* has a very high yield, and is more resistant to diseases than the weaker *Arabica*.

Because it is more bitter, *Robusta* is sometimes over-roasted to get rid of this bitterness.

Robusta stays on the tree after it ripens. So the pickers can visit the *Robusta* coffee tree fewer times than they have to visit the *Arabica* coffee tree, thus keeping the labour costs down.

A CUP OF EACH

The amount of caffeine in your cup varies with the strain of coffee, and the way you serve it. If you start with equal amounts of ground roasted coffee, a cup of *Robusta* will have twice as much caffeine as a cup of *Arabica*. A strong, brewed cup of *Robusta* might have 180 mg of caffeine, while a weak instant cup of *Arabica* might have only 60 mg.

PHARMACOLOGY OF CAFFEINE

Caffeine has the chemical name of *trimethylxanthine*, and the chemical formula $C_8H_{10}O_2N_4.H_2O$, and belongs to the class of chemicals called alkaloids. It is relatively easy to make in a laboratory. It was first discovered in tea in 1827 (and called 'theine'), and then later found in coffee in 1838 (when both it and theine were called 'caffeine').

At room temperature, it takes the form of either silky needles or a white powder, both of which dissolve very easily in hot water. When you cool the solution, crystals of caffeine monohydrate form out of the solution. Pure caffeine does not have any odour, but it does have a bitter taste.

Back in 1981, Solomon Snyder and his co-workers at the Johns Hopkins School of Medicine were the first to explain how coffee stimulated you — on a molecular level. The shape of the caffeine molecule is similar to the shape of a chemical called 'adenosine'. Adenosine is a relatively simple chemical that contains nitrogen. Snyder found that the body has several types of adenosine receptors, which are found in different ratios in different tissues. But caffeine can affect all of these receptors, which is why it affects so many bodily systems.

Caffeine can also stop adenosine from working. Adenosine is a 'relaxing' chemical that works by reducing the firing of nerve cells. So if you stop the natural relaxation of adenosine, you end up stimulated.

Adenosine works by landing on adenosine receptors in your cells — and this tells the cells to slow down. When a caffeine molecule has landed on an adenosine receptor, it stops an adenosine molecule from landing on that same receptor, but it does not trigger the adenosine receptors into action. Caffeine behaves like a 'dud' adenosine. This means that you don't get the 'relaxing' or 'slowed-down' effect of adenosine.

This knowledge about how caffeine affects adenosine receptors could help explain why caffeine is addictive.

One theory is that if you have lots of caffeine in your bloodstream, your body responds by making extra adenosine receptors in the blood vessels to mop up the caffeine molecules. This then allows the adenosine to land on the adenosine receptors.

But when you suddenly reduce your regular daily caffeine dose, the adenosine naturally present in your system lands on all those extra receptors (that were made in response to the caffeine). This causes a sudden relaxing of the blood vessels in your head, which get bigger and give you a headache.

FROM THE TREE TO YOUR MUG

There are about 20 separate steps between the berry on the tree and the hot cup of coffee in your hand. Over half of these steps involve human labour.

The flavour and taste of coffee comes from two sources — the natural chemicals in the unroasted green bean, and the chemicals that are made when the natural chemicals are heated.

The flavour and the taste can drastically change depending upon how you heat it.

The old way of roasting coffee had the coffee beans in a rotating drum that was placed above a source of heat. The new way of roasting coffee involves blasting hot air through the coffee beans, so that they tumble and rotate.

As the coffee gradually heats up, it becomes about 20 per cent lighter (the weight evaporates in the form of steam, carbon dioxide, carbon monoxide, and other chemicals). The beans change colour from green to a deep brown and also *swell*, sometimes doubling in size.

Most importantly, the roasting brings on that magnificent characteristic aroma of coffee. In fact, some manufacturers deliberately roast *Robusta* for a wee bit too long, to get rid of some of *Robusta's* unpleasant flavours.

Beans that have been roasted for a long time have a stronger, and more mellow, flavour than beans which have been lightly roasted.

Generally, 193°C (380°F) gives a light roast, while 205°C (400°F) gives a medium roast. A dark roast usually involves temperatures of around 218°C (425°F). The roasts each have a different name, depending on their colour, such as 'light', 'high' and 'French'.

History of Coffee

There are many contradictory stories about the history of coffee.

We probably began consuming coffee around 500 AD. The earliest written reference to coffee is in a medical manuscript from circa 900 AD. From this time, we had it as a medical drug, and a food, and even added it to wine.

The liquid coffee that we drink today first appeared about 1300 AD. First, the beans were roasted to change them from a green to a brown colour, and to give them more flavour. Second, the brown beans were ground to a powder, then mixed with hot water. At this stage both the dark liquid and the coffee grounds were consumed.

We lived with the bitter taste of coffee for about 300 years. Sugar appeared in coffee in Egypt in the early 1600s, and milk in the late 1600s.

Legend of Kaldi

One very popular story is that coffee was discovered in Ethiopia by an Arab goat herder called Kaldi (or Kaidi), way back in 850 AD. Kaldi saw that after his goats had eaten the red berries of a particular tree, they were very frisky in the daytime and didn't sleep at night. He tried these red berries himself, and found that they had a stimulating effect.

After eating the red berries, Kaldi also began to dance with his goats. By chance, a Muslim holy man saw him dancing around. Kaldi explained to him about the red berries. The legend goes on to say that the Muslim holy man had a vision in his dreams that night. The prophet Muhammad came to him, and told him to boil some of these red berries in water, and give the drink to the faithful who came to his mosque. Muhammad said that the drink would stop the worshippers from falling asleep. It worked. The new drink was so good at keeping people awake, that this mosque became famous for its very lengthy religious services.

Although alcohol is forbidden in the Muslim religion, they called this delicious new drink 'The Wine of Araby'.

Another story about the origin of coffee is that the coffee plants were growing wild on the plains, and that the local tribes chewed the fruits. It was used as both a stimulant and a medicine.

Coffee and Vice

However, soon people found a problem with this new drug. The coffee houses offered more than just coffee — they had singing, dancing and even gambling. So the orthodox priests claimed that coffee was intoxicating, and since the Koran forbade a drug-caused intoxication, they said that coffee should be banned.

The coffee drinkers had luck, and power, on their side. The Caliph of Cairo who was a direct descendant of Muhammad and who also ruled much of the Arab world at that time, was a passionate coffee drinker as well. He wasn't going to give up his favourite drink — so he did not outlaw coffee.

HALF LIFE OF CAFFEINE

Caffeine doesn't seem to stay permanently in the body — it washes in, and then washes out.

When you drink a caffeine-containing drink (coffee, tea or cola), over 95 per cent of the caffeine is absorbed by the gut. It takes a trip to the liver, and then makes its way into the general blood circulation. The highest blood levels of caffeine are reached within 15 to 45 minutes — and then they begin to fall.

The 'Half Life' ($T^{1/2}$) of a drug is the time taken for your body to metabolise, or break down, *half* the amount that you have in your bloodstream. After another $T^{1/2}$, the level of the drug in the bloodstream has dropped to half of its previous level — or 25 per cent of the original level. After the third half life, the drug is down to 12.5 per cent of its peak level.

In humans, the caffeine half life (Caff-$T^{1/2}$) varies from 3.5 to 100 hours, depending on what kind of human you are — male, female, child, etc. In an adult, half the caffeine has gone from your system after five or six hours. But if a woman uses the oral contraceptive pill, her caffeine half life doubles to about 12 hours.

Pregnant women have the same half life as a regular adult, until they get to about three months pregnant. Then from four to nine months, the Caff-$T^{1/2}$ increases to about 10–18 hours. So for a pregnant woman, a little coffee goes a long way. But by one week after birth, the Caff-$T^{1/2}$ is back down to around six hours.

Caffeine can leave the bloodstream of the nursing mother, and transfer to the baby through breast milk, sometimes keeping it awake and inconsolable. The caffeine hangs around in the baby for a *long* time. The half life of caffeine in a premature or a newborn baby is around 100 hours. By the time the babies are eight months old, the Caff-$T^{1/2}$ is down to four hours.

Smokers have a very short Caff-$T^{1/2}$ (3.5 hours). This might explain why smokers tend to drink more coffee than non-smokers. Also, when smokers give up tobacco, they become more sensitive to the effects of caffeine and may need to cut down their coffee intake when they give up smoking. (Poor smokers — having to drop their intake of two drugs at the same time.) To compensate, I suggest an increase in the intake of other 'drugs', such as water and air!

KILLER CAFFEINE

The LD–50 of a drug is the dose that will be lethal to 50 per cent of the population. For caffeine, it's about 200 milligrams per kilogram, which for a typical 70 kilogram male works out to 14 grams.

It's practically impossible to get that much from your regular sources of caffeine (140 cups of coffee, 200 cups of tea, 300 cans of cola in a row). You would begin to vomit before you could drink enough to kill you.

However, in the early 1980s in the USA, several mail-order companies sold pills and capsules that were loaded with big doses of caffeine — up to 500 milligrams (half a gram). Drug dealers bought these legally in large quantities, and then resold them as stimulants to school and university students. In 1980–81, these caffeine tablets killed more than a dozen people in the USA.

It was difficult for the Food and Drug Administration (FDA) to do anything about it because caffeine is a legal drug. The advertisements in the magazines and the pamphlets were worded very carefully to avoid breaking the law. Eventually, the pills were banned because the US Postal Service said that the mail-order companies were misrepresenting the safety of the drugs.

Coffee Travels the World

In the 1200s, coffee plants were carried to Yemen and Turkey, breaking the monopoly that Ethiopia had enjoyed. There was massive coffee-tree cultivation in the 1400s and 1500s.

The Ottoman Turks brought coffee into Constantinople in 1453, with the first coffee house opening there in 1550.

The coffee bean slowly spread into Europe and the Americas in the 1500s and 1600s. The coffee tree was imported into Indonesia — then called Java — in the 1600s.

In 1573, a Dutch explorer called Reuwols brought news to Europe of a strange and delicious new drink called 'coffee'. Around this time, the Egyptians were so worried

about how coffee could damage the social and moral fabric of their society, that they banned it, and burnt all their stocks of coffee.

The Arabs had a monopoly in growing coffee until 1600, when smugglers took 'seven seeds' of unroasted coffee beans from Mocha to Southern India where trees grew from these seeds.

Coffee in England

The word coffee was first mentioned in the English language in the year 1601. William Parry wrote about an adventurer, Anthony Sherley, who brought coffee to London, where he sold it at five pounds per ounce.

Coffee reached Italy by around 1615, and France in 1644.

In the early 1650s, the first coffee house opened in England. (Tea, that other caffeine-containing drink, was introduced into England in 1657.) By 1657, English advertisements claimed that coffee could cure practically anything — including scurvy and gout.

In the 1600s, coffee houses were also called 'Penny Universities'. This was because it cost only a penny to get in, and people from all walks of life would meet to discuss all sorts of issues and ideas — including politics, the law, books and business. But in Turkey, the Grand Vizier thought that these coffee houses could easily turn into hotbeds of revolution, and he ordered them all closed. He was very serious about closing coffee houses. First offenders were very badly beaten. Second offenders were sewn into leather bags and thrown into the Bosporus. King Charles II of England called coffee houses 'Seminaries of Sedition'.

Around 1675, 'The Women's Petition Against Coffee' was published in England. At that time, women were not allowed in the coffee houses. They claimed that this segregation stopped them from spending enough time with their husbands. It was also claimed that the large amounts of coffee drunk were making the men impotent.

HOW TO MAKE COFFEE

There are four main ways to prepare a drink of coffee — brew, percolate, drip filter, and espresso. As we got more knowledgeable about coffee, we prepared the red berries more. We began with the green seeds, moved to the roasted brown seeds, and finally realised around 1300 AD that best of all were the roasted brown seeds after they had been ground. Of course, the fineness of the grind is important.

Brewing is the oldest method. You just boil up the ground coffee, and then filter it. In the early days, they just boiled the red berries. (But boiling makes the flavour more bitter.)

You can also percolate coffee. Percolating involves running boiling water through ground coffee several times, until the liquid gets strong enough. Many percolators are in two separate parts that clip or screw together.

The drip-filter machine is really simple. You just put some ground coffee in filter paper, and run boiling water through it — only once.

Espresso involves forcing boiling water or steam through ground coffee that has been packed tightly. The coffee has to be very finely ground to build up the pressure higher. Espresso has a well-rounded flavour, and is not very bitter because the water goes past the ground coffee very rapidly.

COFFEE ADDICTION

Coffee is probably the most common drug of dependence anywhere on the planet. Babies can be exposed to it *before* they are born, because caffeine can cross the placenta, from the mother to the baby.

Coffee is so addictive, that cutting out just one cup of coffee a day is enough to give some people withdrawal symptoms.

For a long time, surgeons have noticed that some of their patients had very puzzling symptoms after surgery. Sometimes, especially if they have had surgery to the gut, patients do not take anything by mouth for a few days after the surgery — and that includes liquids such as coffee. As well as a severe headache, the symptoms that patients experienced included feeling terrible and achey like having the flu, extreme fatigue, and general depression about the future. Doctors have also noticed slower motor function, and reduced alertness.

Coffee withdrawal can explain these symptoms. The withdrawal symptoms are usually greatest 20 to 48 hours after the last hit of caffeine. Now to get 100 milligrams of caffeine, you need one cup of coffee, or two cups of tea, or three cans of a cola. One study looked at 62 adults, aged between 18 and 50, who drank about two to three cups per day. When their caffeine intake was cut by 100 milligrams a day, half of them had headaches.

The withdrawing coffee addicts also scored more highly on scales of depression and anxiety.

An American doctor, Joseph Weber, treats this 'Post-Surgery Withdrawal Syndrome' by injecting caffeine directly into the blood — 240 milligrams in one hit. This dose can reduce the chance of a headache after surgery, from 25 per cent to 10 per cent. (The intravenous caffeine is strictly for medical purposes — not for your quick afternoon lift.)

If you want to withdraw from your coffee addiction, the easiest method is to cut down your consumption a little bit each day.

Coffee and Conversation

In the second half of the 1700s, London coffee houses became famous for brilliant conversation. One very famous group was The Club. Some of the people involved were Doctor Samuel Johnson and his biographer James Boswell, Oliver Goldsmith (author), David Garrick (actor), Richard Brinsley Sheridan (dramatist) and Sir Joshua Reynolds (painter).

Coffee Conquers Europe

The first coffee house opened in central Europe in 1683. At this time, the Ottoman Turks were trying to invade Vienna with their 300 000-man army. They had brought sacks of strange green beans with them, which they left behind when they retreated. The Viennese were not familiar with coffee. But one Austrian spy told them that they should roast the beans, then grind them up and then boil them in water and filter off the grounds. He opened up his coffee house, Roten Kreuz (At the Red Cross), and suddenly, Vienna fell in love with coffee.

In 1713, King Louis XIV of France was given a living coffee bush. By 1754, Paris had 56 coffee shops.

. . . and then the New World

The descendants of this coffee bush set off the vast coffee industry of the Americas. In 1723, Gabriel Mathieu de Cheu (or Clieu), a French naval officer, broke into the French royal gardens, and stole a plant. He kept it alive, on the long dry journey to Martinique in the Caribbean, by giving it some of his precious drinking water.

HOW LONG TO BREW COFFEE

Every day, millions of people around the world do a chemistry experiment when they pour hot water over roasted ground coffee beans.

But how long should they let the coffee brew?

In 1991, James Hardy and Terrence Lee, chemists from the University of Akron, decided to measure how the brewing time would change the chemicals in the coffee. They let their coffee brew for periods ranging between four and 12 minutes, and immediately chilled each sample to trap the delicate coffee chemicals. Then they used a combined gas chromatograph and mass spectrometer to actually measure 30 separate chemicals from each sample of coffee. (But today we know of about 500 different chemicals in coffee.)

They found that the best brewing time was six minutes.

If the coffee brewed for less than six minutes, the hot water didn't have enough time to extract all the chemicals that make the coffee taste nice. And when the coffee brewed for more than six minutes, many of those nice-tasting chemicals evaporated into the air.

But one problem that Hardy came across was that coffee is not ground to a consistent diameter — the grains all have different sizes. Some of the very finely ground particles will give up their flavours in just a few minutes, but the larger size grains need a longer brewing time. So if the coffee grains are too different in size, it's hard to find a brewing time that suits all of the coffee grains.

HOW TO MAKE DECAF

Decaffeinated coffee began in 1903. A cargo of coffee beans that had accidentally soaked in seawater were delivered to Ludwig Roselius, a German coffee importer. He didn't want to throw them away, so he asked chemists to look at them. They realised that they could remove caffeine from coffee, without destroying too much of the delicate coffee flavours. In French, 'sans caffeine' means 'without coffee', so he called his decaf 'Sanka'.

There are a few different ways to remove the caffeine from the green coffee. Basically, they use a liquid that absorbs the caffeine, and then they remove that liquid. One method is solvent extraction. First, the beans are steamed. They absorb water, and plump up. This brings the caffeine to the surface of the beans. An organic solvent (such as methylene chloride) is then washed over the beans, and takes the caffeine with it. Another method uses liquid carbon dioxide.

Stolen Coffee Bushes

The coffee plants which yielded coffee beans were very tightly guarded, but it was impossible to stop them from being stolen. Over a period of many years, plants were smuggled out of Arabia. The Dutch grew smuggled coffee plants in the botanical gardens in the Netherlands, before they sent them off to Indonesia, from where they spread throughout the tropics.

The British and the French also tried to make money from the coffee tree. France and Britain couldn't grow coffee at home, because the climate was unsuitable. In 1825, the British first planted coffee in Sri Lanka (then called Ceylon).

The coffee tree reached Hawaii in 1825.

By 1842, Vienna had 15 000 coffee shops (but this number had dropped to just 1250 by 1925).

Sri Lanka – The Need for Tea

In 1869, the coffee rust fungus first appeared in the coffee plantations of Sri Lanka. It eventually destroyed the local coffee-growing industry. The shortage of coffee led to a huge increase in coffee prices which, in turn, caused a sudden surge in tea growing and drinking.

By 1884, Sri Lanka's annual coffee export dropped from a peak of 700 000 bags to 150 000 bags. The last shipment of coffee beans left Sri Lanka in 1899.

Some people say that if this rust had not attacked the coffee crops of Sri Lanka, tea would never have become as popular as it is today.

FILTER COFFEE

A major step in the search for the perfect cup of coffee came in 1940 with the cone drip filter system. Dr Peter Schlumbohm migrated to America in 1939 from Germany. In 1940, he came up with an incredibly simple, elegant and effective design. He used a common piece of scientific laboratory equipment — a Pyrex glass container. The amazing thing about Pyrex is that while it is a perfectly transparent glass, it doesn't crack when exposed to sudden changes in temperature. You can safely take it from the fridge to the hotplate.

He placed an upside-down cone with holes punched in the tip, on top of this simple Pyrex container. He put a conical paper filter in it, added some ground coffee, and then poured in freshly boiled water. The water absorbed the colour and flavour from the ground coffee, went through the paper and into the Pyrex container, leaving the grounds behind.

He had difficulty in selling such a simple and elegant design, because it looked too simple and obvious. Eventually he was able to persuade a buyer from Macy's in New York City to take one of his Chemex coffee makers home. Before midday the next day, he had a phone order for 100 Chemex coffee makers.

But there was a problem — World War II.

Pyrex was made by Corning Glass, and it had a special wartime military classification. Civilians were not really supposed to have access to large quantities of Pyrex. In addition, Corning Glass were not allowed, under wartime regulations, to make a new product that was not related to the war effort, without specific clearance from the War Production Board.

So Schlumbohm wrote a letter directly to President Roosevelt.

In his letter, he wrote two very important Latin sentences. The first one was an old Latin proverb — *'minima rex non curat'*, which means *'a king does not bother with minor details'*. Then he added another Latin sentence that he invented himself. It was *'Sed President curat et minima'* which means *'but a President cares even about details'*. Luckily, President Roosevelt loved coffee, and Chemex got its priority clearance. By 1941, Chemex coffee makers were on the market. But then coffee rationing began in November 1942.

Australian Coffee

In Australia, Cairns in far northern Queensland was producing coffee from the 1890s. By the year 1900, 40 per cent of all of Queensland's coffee needs were supplied internally.

CAFFEINE - GOOD MEDICINE

In small doses, caffeine will speed up your general metabolism. It will also increase your body temperature, as well as your alertness. Caffeine can mobilise your body fat, so that exercising muscles can use it as fuel. It also stimulates the secretion of gastric acid, which supposedly helps you digest your meal better.

Olympic-grade athletes like another of the effects that it has on muscles — it's a 'muscle twitch potentiator'. Perhaps this is how it speeds up your reactions.

Caffeine stimulates and enhances your ability to do intellectual tasks.

Caffeine also relaxes your airways. This makes it easier for air to flow in and out of the lungs. Caffeine increases your breathing rate, and the adrenaline level in your bloodstream — athletes enjoy these effects.

Caffeine also closes down blood vessels in the brain, which is why it can relieve headaches. However, it opens up the blood vessels in the extremities, and around the heart — once again, good for athletes. Caffeine has a direct effect on how fast your heart beats, and how much blood it squirts out with each stroke (usually about 80 ml).

Coffee can help you work harder, and at the same time, not overload your heart. One study by the Oregon Health Science University looked at six (not a very big sample size!) healthy young men. They worked on exercise bicycles both before, and after, a double espresso. After they had drunk the double espresso, they could pump more blood through their body at a lower blood pressure and, at the same time, use less oxygen to do so. This was probably because of the vasodilator (opening up the blood vessels) effect of caffeine.

When caffeine is combined with other pain-killers, the resulting combination works better than either caffeine or the pain-killer by itself. Caffeine also increases the anti-cancer action of certain drugs.

Coffee Houses

Coffee houses again took on their early role as 'Seminaries of Sedition' in the West, in the 1950s and 1960s. First the beatniks, and then the hippies, exchanged ideas about books and politics, sometimes on a background of folk singing.

What will the next wave of coffee-induced 'sedition' bring?

That Mystery Coffee Ring

Millions of people now start their day with coffee, and sometimes spill it — which leads us back to the mystery of why all the dark colour ends up at the outer rim of the coffee stain.

This mystery was solved by a Chicago scientist.

CAFFEINE - BAD MEDICINE

Caffeine can cause anxiety and tension, panic attacks, insomnia, headaches, diarrhoea, loss of appetite, irritability, dizziness, twitching muscles, hot flushes, upper abdominal pain, a rapid and irregular heartbeat, and a decrease in hand steadiness. Some surgeons who do very delicate work cannot have coffee before an operation.

It also stimulates the secretion of gastric acid, which is why it is unhealthy for people with gastric ulcers. Coffee is also known to help cause gastric reflux (or gastric regurgitation), where the contents of the stomach come back up your oesophagus. It does this by weakening the lower oesophageal sphincter at the junction of the oesophagus and the stomach.

If you have some caffeine before you go to bed, it will delay and shorten the amount of sleep that you have, and will also reduce your deep sleep cycle. While it will reduce the amount of dream sleep overall, it will increase it early in the night.

Most people find that coffee does not interfere with their sleep — until they hit middle age. They may have drunk coffee as an after-dinner beverage for 20 years and had no problems sleeping, and then will suddenly find themselves experiencing insomnia.

Some people can become so sensitive to coffee, that a single cup can give them tinnitus (ringing in the ears), palpitations of the heart, restlessness, and insomnia.

Caffeine also acts as a diuretic, increasing your production of urine. This could lead to dehydration.

Sidney Nagel, a scientist at the Materials Center at the University of Chicago, treats his morning coffee differently from most other people — he prefers to spill it. On one particular shaky morning, he stared at the dried coffee stain on his kitchen bench. It was much the same as thousands of other dried coffee stains he had seen in his life. But on this particular morning he asked himself, why did it have a dark rim? Why wasn't the coffee stain an even colour all over?

When he got to work that morning, he asked all his fellow scientists why a coffee stain was darker on the outside rim, and nobody knew why. He realised he had stumbled across a fundamental question.

Scientific Coffee Spilling

In science, theory and experiments leapfrog each other, so the scientists at the Materials Center began experimenting by spilling coffee on every flat surface they could find.

They found a few surprises. Even if they spilt coffee onto a surface and then turned it upside down, the evaporating coffee would leave behind a dark rim.

THE WORD 'COFFEE'

Where did the word 'coffee' come from?

The Arab word 'qahwah' means 'wine' (as in coffee is the Wine of Araby). There is also the province, 'Kaffa', in Ethiopia, where coffee was first known to be grown. The Turkish word, 'kahve', is similar to both 'qahwah' and 'Kaffa'.

Our modern word 'coffee' comes from 'caffé' the Italian version of 'kahve'.

In turn, the word 'coffee' gave us the word 'café', a place where coffee is drunk.

The rim appeared with other *liquids* besides water, such as various oils and alcohols. It appeared with other *substances* besides coffee, such as salt and polystyrene colloids. And the rim appeared on virtually any *surface*, including glass, metal and plastic.

Clue 1 - Evaporation

Olgica Bakajin, one of the students, then did a very clever experiment. She covered different parts of the liquid, so it couldn't evaporate. She made one cover as a doughnut shape (with a hole in the middle), so the inside of the coffee blob was exposed, while the outside of the coffee blob was covered. The liquid coffee inside the doughnut (exposed to the air) could evaporate, but the outside coffee (under a cover) could not. Under the cover, where the coffee could not rapidly evaporate, there was *no* dark stain. And in the exposed area, there was no well-defined ring, just a vague blemish. But then she cut away part of the cover, and exposed a small segment of the rim of liquid. Now the coffee could evaporate from that section, and a dark ring developed where there was no cover.

This experiment told her that evaporation was somehow involved in the thrill of the spill.

Clue 2 - Migration

Bakajin and another student, Robert Deegan, dumped thousands of microscopic

COFFEE-INSPIRED INSURANCE SALESMEN

Lloyd's of London came into existence because Edward Lloyd opened a coffee house in Tower Street in 1688. Tower Street was close to the docks of London. Bankers and merchants, as well as seafarers and insurers, all gathered there to carry out business in a friendly and fragrant atmosphere. Soon, the biggest marine insurance business in the world blossomed into existence — Lloyd's of London.

INSTANT COFFEE

It took until the 1950s, before instant coffee really took off as a commercial product. It's often made from the cheaper *Robusta* variety.

The man who 'invented' instant coffee (also called soluble coffee in the trade) was George Constant Louis Washington, an engineer. He stumbled across instant coffee in a mountain range in Guatemala in 1906 (or 1907). He came across a coffeepot that had a fine powdery brown residue around the mouth, where it had boiled over with coffee. He tasted this brown residue, and was surprised at how pleasant the flavour was.

Previous attempts to make instant coffee had been carried out at around sea level, where water boils at 100°C. The residue left behind from boiling down the coffee tasted terrible. But this coffee tasted better, because the coffee had not been heated to 100°C. (As you go up to higher altitudes, the boiling point drops. The coffee in the coffeepot had boiled at a very high altitude, where the boiling point is about 85°C.)

He used this principle, of 'low temperature boiling under reduced pressure', to open the G. Washington Coffee Refining Company in Brooklyn in New York in 1909. In World War I the United States Army made instant coffee part of the standard army rations.

There are three main ways to make the instant coffee powder or granules. First, the concentrated liquid coffee is squirted at high pressure through a small hole, which turns it into a spray made up of very small particles. When this spray is blown through a blast of hot air, the water evaporates almost instantly, leaving behind the powder.

The second method is to heat the liquid under very low pressure, or a vacuum. Once again, as the liquid evaporates, a powder is left behind.

The third method is to freeze the concentrate into blocks, which are then broken up into much smaller particles. These granules are then put into a vacuum chamber, where they are heated gently. Under low temperature and pressure, the water in the ice turns directly into vapour without going through the liquid phase and evaporates off, leaving much of the original coffee flavour behind.

polystyrene spheres (1 micron in diameter = one millionth of a metre) into the coffee. They then looked with a microscope at the coffee as it evaporated. Now if you look at any liquid under a microscope, you'll see a phenomenon called Brownian Motion, where all the particles in the liquid jiggle this-way-and-that randomly. (Einstein thought about Brownian Motion for a while, and you can find Brownian Motion in the movements of prices on the stock market.)

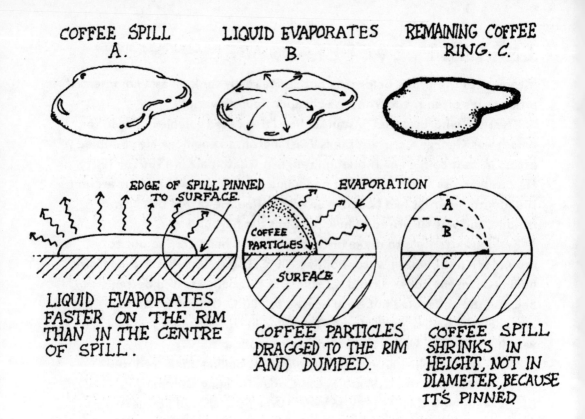

COFFEE SPILL
A.

LIQUID EVAPORATES
B.

REMAINING COFFEE
RING. C.

EDGE OF SPILL PINNED
TO SURFACE.

EVAPORATION

COFFEE
PARTICLES ↓

SURFACE

A
B
C

LIQUID EVAPORATES
FASTER ON THE RIM
THAN IN THE CENTRE
OF SPILL.

COFFEE PARTICLES
DRAGGED TO THE RIM
AND DUMPED.

COFFEE SPILL
SHRINKS IN
HEIGHT, NOT IN
DIAMETER, BECAUSE
IT'S PINNED

But they did not see totally random motion. As the coffee stain evaporated, they saw the tiny spheres definitely moving outwards, towards the edges of the spilt liquid.

Answer – Pinned Rim

As they continued with their experiments, they saw that as the liquid in the spilt coffee evaporated, the spill did not shrink in diameter — it shrank in height. They realised that tiny rough patches on the surface were somehow 'pinning' the edge of the liquid — stopping it from shrinking towards the centre as the liquid evaporated.

Evaporation was another factor. The liquid in the centre of the spill could evaporate only upwards. But the liquid at the edge could evaporate both upwards, and out to the side.

So there was more evaporation occurring at the edge of the liquid, than in the centre. As the liquid at the edges evaporated, more liquid moved from the centre to the outside rim. The outside rim could not move inward to the centre, because it was 'pinned' by the initial roughness on the surface. And as more of the solid stuff in the coffee moved towards the edge, and then got laid down by evaporation, it made the surface even rougher — pinning the edges even more effectively.

So the key factor is the pinning — the fact that as the liquid evaporates, it doesn't shrink towards the centre, but shrinks only in height. And the second factor is that it evaporates more from the edges (because of the greater surface area available) than from the centre. These two factors combined mean that the liquid flows from the centre to the outside rim.

COFFEE - GOOD FOR PREMATURE BABIES

Premature babies often have problems with their lungs. But coffee can help.

One of the main problems that premature babies suffer is 'apnoea' ('a' means 'not', and 'pneu' means 'breath'). Apnoea means that they stop breathing for periods of 20 seconds or more. This apnoea can happen as frequently as once every hour. Apnoea also affects about 85 per cent of babies weighing one kilo or less.

This fact is hard to believe, but the trigger for humans to breathe is not low oxygen — it's high carbon dioxide. In other words, the main reason that you breathe is that you have high levels of carbon dioxide in your blood, and a receptor senses this, and makes you suck more air *in*, so you can blow the carbon dioxide *out* of your lungs.

The immature nervous system of the premature baby often does not respond to high levels of carbon dioxide in the blood — so the baby stops breathing. When the baby stops breathing, the oxygen levels in the blood drop. Sometimes they can drop so dangerously low, that they cause permanent damage to the hearing and to the brain.

One of the drugs commonly used to treat apnoea, aminophylline, has side effects which include damage to the heart and nerves.

A team from the **University of Queensland** has found that caffeine is a safe alternative to aminophylline for premature babies, who have arrived up to 15 weeks early (at 25 weeks, instead of 40 weeks).

FAMOUS COFFEE DRINKERS

Part of the reputation of coffee is that it is supposed to stimulate intellectual activity. A few intellectual heavies really liked their coffee. Beethoven was supposed to drink a very strong brew — 60 beans in every cup. Voltaire, the French philosopher, was rumoured to have drunk 72 cups a day. And Honoré de Balzac, the French author, drank some 50 000 cups of coffee in his life. The coffee was cold, black, and as thick as soup. He said that it kept him awake during the night, so he could keep on writing.

HEALTH – CAFFEINATED VS DECAFFEINATED

There have been many statistical studies trying to find differences between the health of decaffeinated coffee drinkers and regular coffee drinkers.

It turns out that people who drink a lot of caffeinated coffee are also more likely to be high achievers, to eat junk food and drink alcohol, as well as smoke cigarettes. On the other hand, decaf drinkers do more exercise, eat more vegetables, eat less fat and have lower levels of body fat.

Statisticians call these differences 'confounding factors'. Does drinking decaf make you exercise more, or does having more exercise make you drink decaf, or is there a separate cause that makes you both do exercise *and* drink decaf? These confounding factors have made it very hard to work out how moderate coffee drinking affects your health.

Here's another confounding factor to make it harder to work out the health effects of decaf. Most normal (with caffeine) coffee comes from the low-caffeine *Arabica* bean, while most decaf comes from the high-caffeine *Robusta* bean.

How to Make a Ring

You need only three conditions to get rings happening.

First, the liquid should kiss the surface at an angle that is not zero. A right angle would do very nicely. Second, the outside rim of the liquid should be pinned in position. One very good way to do this is to spill it on a rough surface. Third, you need evaporation.

You'll see this 'Ring Effect' with cola drinks, wines and even clean water.

Useful Theory

It turns out there are actually a few practical uses for the work done by Nagel and his fellow scientists. For one thing, they now have a theory which very accurately predicts the shape and the thickness of the resulting ring. So they could use this knowledge to accurately paint fine gold wires in electronic microcircuits. And this theory could also give us more accurate printing of ink on a page, and help us understand industrial washing and coating processes.

And as far as Nagel is concerned, his life is enormously enriched, and made easier. He no longer feels guilty about grubby kitchen surfaces, and he gets enormous excitement from watching paint dry.

In an article in the *New Scientist*, Nagel said, 'I find watching paint dry very exciting these days'. And he no longer has a super-clean kitchen: 'The rings are so beautiful I just leave them. There are times when I've made a particularly splendid mess and have left it there for a week.'

COFFEE AND MONEY

The world's single most valuable agricultural commodity is the coffee bean. As coffee was introduced into various poor tropical countries around the world, it made (and later broke) their economies.

Today, about 60 to 70 per cent of the world's coffee comes from Latin America and Africa. The largest importer of coffee is the USA. The USA consumes one-third of the world's coffee, which costs it about $1.5 billion per year.

Around 1900 in Brazil, the export of coffee made Brazil wealthy. In 1908, it accounted for 53 per cent of the export earnings of Brazil. In fact, the first few presidents of Brazil were called Coffee Presidents.

In 1931, the world coffee market collapsed and brought economic disaster to Brazil, as well as a revolt in its southern provinces.

Certain groups have made enormous profits by 'controlling' the price of coffee. The coffee crop can fluctuate wildly, depending on whether there is a glut, or a shortage. For example, sudden cold snaps in Brazil (in 1975 and 1994), forced coffee prices upward and kept them there for about three years.

REFERENCES

T. Adler, 'Coffee Can Give Many Species a Boost', *Science News*, Vol. 150, 31 August 1996, p 132.

Phillip Ball, 'How Coffee Leaves its Mark', *Nature*, Vol. 389, 23 October 1997, p 788.

Julian Cribb, 'Caffeine Jolt Gives Early-bird Babies a Better Start to their Days', *The Australian*, 26 April 1996, p 3.

Robert Deegan *et al*, 'Capillary Flow as the Cause of Ring Stains From Dried Liquid Drops', *Nature*, Vol. 389, 23 October 1997, pp 827–829.

Denise Grady, 'Don't Get Jittery Over Caffeine', *Discover*, July 1986, pp 73–79.

John R. Hughes, 'Clinical Importance of Caffeine Withdrawal', *New England Journal of Medicine*, 15 October 1992, pp 1160–1161.

Kenneth Silverman *et al*, 'Withdrawal Syndrome After the Double-Blind Cessation of Caffeine Consumption', *New England Journal of Medicine*, 15 October 1992, pp 1109–1114.

Marijke van Dusseldorp, 'Cholesterol-Raising Factor from Boiled Coffee Does Not Pass a Paper Filter', *Arteriosclerosis and Thrombosis*, Vol. 11, No. 3, May/June 1991, pp 586–593.

Van Nostrand's Scientific Encyclopaedia, 1989, Van Nostrand Reinhold, pp 690–692.

Gabrielle Walker, 'The Thrill of the Spill', *New Scientist*, No. 2105, 25 October 1997, pp 34–35.

http://mrsec.uchicago.edu/MRSEC/Nuggets/Coffee/

MAGGOTS GIVE LIFE

Flies have never had a very good public image. But for a few centuries at least, healers have used flies to cure diseases. And over the last decade, this disease-fighting aspect of flies has been rediscovered.

Flies in History

In general, people hate flies. This dislike goes deeper than simply being annoyed by flies buzzing around your head.

TV ads for flyspray talk about flies 'spreading disease with the greatest of ease'. The Bible says that the house of the Pharaoh was plagued by 'a grievous swarm of flies'. And the Israelites had a rather insulting nickname for the Chaldean God Baal — they called him Beelzebub, which means 'Lord of the Flies'. (This name probably relates to the thousands of flies that buzzed around the sacrificed animals in the Chaldean temples.)

This dislike of flies has been linked to disease throughout history. Moses told the Israelites to bury their excreta, to keep flies away from human wastes. The Greeks and the Romans speculated that perhaps flies could spread dysentery by landing on food. Back in 1498, Bishop Knud of Denmark claimed that one of the early signs of the approach of the plague was an increase in the number of flies. And Thomas Sydenham, the famous English physician of the 17th century, said that, 'if swarms of insects, especially houseflies, were abundant in the summer, the succeeding autumn was unhealthy'.

Maggots are 'baby' flies, and in the 1600s, people thought that having maggots in your brain would make you mad. Way back in 1620, John Fletcher, in his play *Women Pleased*, wrote:

Are you not mad, my friend? What time o'th moon is't?

Have not you maggots in your brain?

Flies Cause Disease

Today we know that flies carry over 100 different types of nasty germs. They certainly are known to cause cholera, dysentery, conjunctivitis and trachoma. It has been claimed that they can spread more than 65 diseases, including tuberculosis, Coxsackie virus infection, yaws, leprosy, polio and typhoid. But the fact that flies *carry* these germs (bacteria, viruses, parasites), does not mean that these germs actually infect humans, and cause diseases. In some cases, the germs don't cause any problems at all. Flies are innocent of some of the accusations against them.

But using flies to *cure* diseases — well, that's a whole different way of thinking. When we use flies to cure diseases, it's not the *flying* version of a fly we use, but the *wriggling* version — the maggot.

Fly Changes Shape

The fly goes through a complete change of shape (metamorphosis) in its life cycle. The different stages are fly, egg, larva (sometimes called a maggot, depending on the fly), pupa, and then back to fly again.

The housefly lays its eggs, and within a day, these eggs hatch into little wriggling maggots. These maggots are the larval stage of the fly. The maggots grow quite rapidly, having two moults (shedding their old skin) along the way. At around day five, the maggot stops eating, and shrinks a little before it turns into a pupa. Maggots spend another week-and-a-bit in this hardened cocoon-like stage, before they emerge as cute little baby flies, ready to start the cycle again.

HOW TO SWAT A FLY

Dr Rene Bult, a scientist from Groningen University in the Netherlands, has been studying flies for five years. He claims to have discovered the secret of how to swat a fly successfully — wear red clothes, use a red fly swat, and do it in the afternoon.

Why? First, the housefly's favourite colour is red. Second, it gets tired in the afternoon. It uses about 75 per cent of its brain power to run its sense of vision. So in the afternoon, it has enough of its sense of vision remaining to still see red (and be attracted), but not enough to avoid a fast-moving fly swat.

LARVAE — BIOLOGY

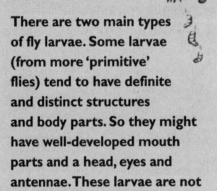

There are two main types of fly larvae. Some larvae (from more 'primitive' flies) tend to have definite and distinct structures and body parts. So they might have well-developed mouth parts and a head, eyes and antennae. These larvae are not usually called maggots.

But some (from more 'recent' flies) don't have eyes, wings, legs, antennae or most other well-defined body parts. They have one or two mouth hooks, instead of a complicated head. It is these larvae that are commonly called maggots. At the end of their time as a larva, the internal tissues turn into a creamy mush, that somehow turns into wings, legs, and so on.

Can Flies Cure Disease?

People have known about the healing power of maggots for ages. The Maya civilisation used them 1000 years ago.

The military have long known the value of maggots in fighting certain diseases. It's always been important for military doctors to know about infectious diseases. In practically all wars in history, until the Korean War, infectious diseases killed more soldiers than the weapons of the enemy.

In the 16th century, doctors on the battlefields of Europe noted that if a soldier had a wound that was infested with maggots, that particular wound usually healed quite well.

The surgeons who travelled with Napoleon's armies knew that wounds infested with maggots recovered much better than the wounds which did not have maggots crawling through them. And during the American Civil War, army surgeons deliberately put blowfly maggots into infected wounds to clean away infected and diseased tissue.

In World War I, William Baer, an American doctor, happened to be treating two soldiers who had lain, unattended, on a battlefield for a week. Their belly wounds had become infested with thousands of maggots. But these soldiers recovered much better than soldiers who'd been treated in the military hospital. After the war, at Johns Hopkins University in Baltimore, Maryland, Baer examined this 'Maggot Therapy' further. He showed that maggots could cure some very tough infections. By the early 1930s, over 300 hospitals were using Maggot Therapy. But during the 1940s when the first true antibiotics, the sulfadrugs, were invented, Maggot Therapy was quickly forgotten.

Interest in maggots re-emerged in 1982 when an orthopaedic surgeon, John Church, was called to treat a car-crash victim who had lain unconscious for three days, at

MAGGOTS MAKE DWARVES

In ancient Scandinavian (Norse) mythology, dwarves were clever workers who made many of the riches of the gods. They were usually quite amicable to people. One Norse legend claims that when the giant, Ymir, was killed, maggots appeared in his skull and came out his eye cavities, then supposedly turned into dwarves.

the bottom of a deep ditch at the side of the road. The young man had deep lacerations on his face and body. He also had massive infestations of maggots crawling all over these wounds. The usual surgical treatment in cases like these was to slice away the infected dead tissue while the patient was under a general anaesthetic. Mr Church was amazed to see that underneath the crawling maggots, the man's wounds were so clean that they had already begun to heal.

MAGGOTS IN FOOD

In the USA, the FDA (Food and Drug Administration) sets limits for 'natural or unavoidable' contaminants in some foods. For example, mushrooms are allowed to have 20 maggots in each 100 gram can, while tomatoes are restricted to only two maggots per 500 gram can.

These maggots cause humans no harm (they have been killed by the processing). In fact, if we accepted slightly higher levels of maggots in processed foods, farmers could use fewer pesticides!

HOW MANY MAGGOTS?

Another team looking at Maggot Therapy consists of John Church (orthopaedic surgeon), David Rogers (an Oxford University entomologist, or insect scientist), and Paul Embden (another entomologist). According to their studies, you need around 10 maggots living on each square centimetre of open sore, to do a decent job of healing the wound.

How Maggots Cure

So, how does Maggot Therapy work?

First, we know that the maggots eat only the diseased flesh, leaving behind the healthy flesh.

Second, they excrete chemicals which kill some of the bacteria that they don't swallow. And third (and this is a rather subtle effect), they are thought to give the living flesh a gentle and therapeutic massage as they crawl over it.

We all know that bandages need to be changed regularly. In a similar way, most patients need at least three changes of maggots!

Maggots can do things that standard antibiotics can't. So they can work in some situations where antibiotics are useless. Some strains of bacteria are becoming resistant to antibiotics. But these bacteria have no defence against maggots physically eating them. Also, when a patient has a vastly weakened immune system, which can't help the antibiotics wipe out the germs, maggots can do a fine job on their own.

Finally, if the body part that is infected has a very poor blood supply, the antibiotics don't get carried to where the infection is. But if you put the maggots where the pus is, they'll eat it up. You can't always get the maggots to the wound, though, for example, if it is an internal wound.

You have to be careful to get the right kind of maggot in your wound. Some maggots, such as the screwworm fly maggot, will eat living flesh. So you don't use screwworm fly maggots. Blowfly maggots are fine because they eat only dead flesh.

SUPERSONIC FLIES

Back in 1926, an American scientist claimed that the deer botfly (*Cephenemyia pratti*) could fly at 1316 kph (818 mph) at a height of 12 000 ft! This is greater than the speed of sound. This claim was complete madness.

First, back in 1926, how could you *measure* such a speed at such a height? After all, there were no supersonic planes, radar, etc. Second, the fly would have needed to eat 1.5 times its own weight every second to generate the 1.1 kWatts (1.5 horsepower) needed to push it through the air at that speed. Third, the increased temperature caused by air friction would have killed it. (Your average passenger jet at a cruising speed of about 800 kph (500 mph) has its skin warmed by about 20–25°C by air friction alone. The heating effect is greater at supersonic speeds.)

Actually, the fastest insect is probably the Australian dragonfly (*Austrophlebia costalis*). It can briefly reach speeds of 58 kph (36 mph).

Modern Maggot Therapy

Ronald Sherman, from the Veterans Affairs Medical Center at Long Beach in California, has been studying Maggot Therapy. He has been working with patients who have injuries to their spinal cord and who, as a result, are partly paralysed. Because they are unable to move in their bed, they sometimes get a rather nasty injury called a 'pressure ulcer'.

Dr Sherman's study examined 10 patients who received Maggot Therapy for their

GOOD FLIES

Maggots and flies are useful in nature, because they help break down dead animals, and recycle and redistribute the chemicals from which the animals were made. The larvae of flies are also a good food supply for other creatures.

pressure ulcers. He found that the maggots healed the wound more rapidly than any other non-surgical method, and that this was a safe, simple and inexpensive therapy with no side effects. (Mind you, in this study the sample size of just 10 patients is too small to be statistically significant. We need more studies with bigger numbers.)

In Dr Sherman's experiment, the maggots made the wound shrink by about

PROBLEMS WITH MAGGOT THERAPY

There are two problems with this Maggot Therapy.

There is the 'yuck factor' associated with having maggots crawling over your wound.

The second issue is that maggots crawling all over a wound goes against the Sepsis Theory, which states that cleanliness is essential for good medical care — and maggots definitely do not look clean and are associated with disease.

25 per cent each week. Most of the wounds healed within a week, and all the wounds had healed totally within a month.

Dr Sherman doesn't just wait for flies to come and lay their eggs in a patient's wounds. First, he hatches his maggots, and then washes them in a chemical solution to stop them from carrying unwanted germs to the patient. Second, he makes a kind of 'maggot pack' with glue, gauze and tiny one-mm maggots and applies this maggot pack to the wound. Then, when the maggots have grown to about one centimetre long (which takes about two or three days), he peels off the maggot pack. The patients sometimes report feeling a 'gnawing or scratching' in the wound, but it doesn't really bother them.

Currently, the maggots are mainly being used when all other medical remedies have failed. It would be interesting to see how well they worked if they were used earlier. With antibiotics becoming less effective, we may be on the edge of a fully blown, fly-blown, medical renaissance.

REFERENCES

Bernard Greenberg, 'Flies and Disease', *Scientific American*, July 1965, pp 92–99.

——'Maggoty Cure for Injured Soldiers', *New Scientist*, No. 1572, 6 August 1987, p 31.

Tracy L. Pipp, 'Take Two Maggots and Call us in the Morning', *Detroit News*, 7 November 1997.

Lois Rogers, 'Doctors Find a Friend in Maggots', *The Australian*, 23 January 1995, p 6.

——'Scientists Breed Maggots for Therapy', *USA Today* (Health), 9 October 1997.

R.A. Sherman, F. Wyle and M. Vulpe, 'Maggot Therapy for Treating Pressure Ulcers in Spinal Cord Injury Patients', *Journal of Spinal Cord Injury*, Vol. 18, No. 2, April 1995, pp 71–74.

Carl Zimmer, 'The Healing Power of Maggots', *Discover*, August 1993, p 17.

MAGGOTS DATE DEATH

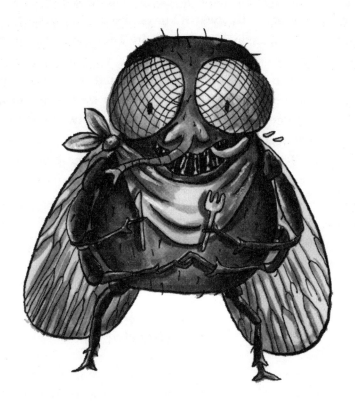

There are a few sure-fire stock-in-trade formulas for TV shows and 'pulp fiction', and one of them is the murder mystery. As part of the investigation into the murder, the police need to know when the victim died. It turns out that the police have a new ally — insects.

LARVAE CAUSE 'MYIASIS'

When fly larvae eat living tissue, they are usually harmful to the creature that they have invaded. They cause an infestation called a 'myiasis'. In myiasis, the results can range from harmless colonisation, to death.

Different species of larvae attack different creatures. The larvae of Hessian flies attack wheat. Other larvae attack oats, onions, cabbage and barley.

In Australia, sheep blowflies (the screwworm fly, *Lucilia cuprina*) do terrible things to living sheep. The female screwworm fly dumps about 300 of her eggs near an open wound. The eggs turn into larvae, which then burrow into the living tissue. The cost to the Australian sheep industry is about $200 million per year.

In Africa, the tumbu fly lays eggs in clothes or sand. The eggs turn into larvae, which hatch once they are in contact with a warm human body. The larvae then dig into the skin, and make boils. They come out of the boils about nine days later.

You can diagnose this particular myiasis by covering the boil with water, and watching for the bubbles. You can also coat the boil with petroleum jelly, which begins to suffocate the larvae, and forces them to come out.

Another fly, the horse botfly (*Gasterophilus intestinalis*) lays its eggs on the sides of a horse. Human horse riders can get infected when they rub their exposed legs against the sides. These maggots have to be removed with a needle.

Insects Finger Murderer

One of the earliest records of how insects helped solve a crime comes from China, in the year 1235, in a small village. The victim had been murdered, and the murder weapon was a rice sickle normally used to cut the rice. The mayor got the men of the village to put their apparently clean sickles on the ground in front of him. Flies landed on one, and only one, of the sickles. They were attracted by the protein from the blood of the murdered man. Even though the sickle had been wiped clean, a thin layer of protein remained. The mother fly needed that protein to provide nutrition for her babies. The owner of that sickle confessed to the murder.

Today, the field of using insects to help solve crimes is called Forensic Entomology. There are only 20 forensic entomologists in the USA, and only one of them is full time. The rest work at universities or museums, and are called in only to help out with crime detection.

The field of forensic entomology really began as a science in 1971. Today, Bill Bass is the official Forensic Anthropologist for the State of Tennessee. But in 1971, Bill Bass was an anthropologist who was mainly interested in studying thousand-year-old bones of American Indians. But from time to time, various police authorities would ask him about a body found in the woods, because they wanted to know when the victim had died. Bill didn't really know, and he soon discovered that there was barely any research in this field, so he set out to build up a database.

The Body Farm

He wanted to see what happened to dead

MAGGOTS - LIFE FROM NON-LIFE

A careful look at maggots sets us on the path to understanding life.

But in 1668, Francesco Redi, a doctor from Florence in Italy, disproved this theory of 'Biogenesis'. Redi was quite famous. Not only was he the personal doctor of the Grand Duke of Tuscany, but he was also a much-admired poet.

Redi did a very simple, elegant and clever experiment. During a period of hot weather, he put meat in several jars. Some of the jars were open to the air, while others had fine gauze over the mouth. The gauze stopped the flies from actually touching the meat.

The meat in the open jars soon began to decay, and was rapidly covered with maggots. The meat in the jars that were covered with a fine gauze decomposed, but did not 'generate' any maggots — because the gauze stopped the flies from laying their eggs on the meat. However, blowflies did swarm over the fine gauze, and laid their eggs on it. Soon, the gauze was covered with maggots.

So Redi proved that the maggots came from eggs laid by flies. Maggots did not come from 'nothing'.

bodies, as time passed by, so he set up 'The Body Farm'. This 1.6 hectares of graveyard and laboratory is owned by the University of Tennessee. It is protected by a razor-wire fence around it and the only entrance is through a door marked 'Biohazard'. Over the last few decades, Bill Bass has scattered some 800 bodies across 'The Body Farm', to see how rapidly or slowly they decay. These bodies have been donated to science. Some bodies are placed in the open fields while others are under trees, or in the front seat of a car or in the boot. Some are in broad sunlight, while others are buried a few metres underground.

By careful observation, Bill Bass has established his database which records what insects do in dead bodies. Back in 1971, investigators at the scene of a murder would ignore flies and maggots on bodies, and view them simply as annoying creatures. But now they see insects as their allies.

Insects and a Corpse

Within minutes after a person's death, a fly will show up, either to eat some protein-rich liquid, or to lay its eggs. As a body decays, it emits various chemicals with names such as 'cadaverine' and 'putrescine'. These chemicals float away from the body and can be smelled by a fly from over a kilometre away. As the ratio of the various different chemicals changes after death, different waves of insects are attracted and come to invade the body (the technical term for this is 'faunal succession'). Initially, flies arrive, then come the ants — to eat the fly eggs. Beetles will eat the fly eggs after they have turned into maggots, and sometimes wasps will lay eggs in the maggot mass. Carrion beetles will chew off some flesh, inject an egg into it, and then bury it underground. Finally, common clothes moths and beetles will strip off everything that isn't bone.

The maggots make strange things happen to the body. Their main aim is to eat the flesh, to grow big and strong, and to keep the life cycle of flies continuing. Around the second week, if you accidentally stumble against the corpse, all the maggots will move at the same time, and the body will appear to jump. Around day 19 after the maggots hatch, they will all leave the corpse, which can quite rapidly lose up to 90 per cent of its weight! (The technical term for this is 'exodus'.)

So in theory, if you know the timetable of the to-ing and fro-ing of the different insects, you can establish the time of death. But this field of forensic entomology is a complicated one. For example, drugs in the body of the victim will change the rate at which different waves of insects are attracted to the body. As a further complication, the insects found in cities are different from those that live in rural areas. The insects in any given area vary at different times of the year, and also change as the temperature fluctuates.

If they have collected the insect samples properly, and noted the weather conditions, the amount of shade on a body and so on, the forensic entomologists can tell the time of death to within six hours or so.

It is too expensive for each police department in the USA to have its own forensic entomologist, so a few places, such as the University of Pennsylvania, run classes to teach police officers about forensic entomology.

So now there's a new position to aspire to for young police cadets — not Detective Chief Inspector, but Detective Chief Insector!

FLIES AFTER DEATH

Flies will sometimes lay their eggs between the lips and the eyelids of people in their dying moments, without even waiting for them to die. Some blowflies actually produce larvae which are live, so this means that within the first few hours of death, tiny maggots can be seen moving on the lips and eyelids. But most of the time, blowflies lay eggs that look like yellow-white grated cheese. They lay these eggs in any of the body openings they can find.

REFERENCES

The Australian Encyclopaedia, Australian Geographic Society, 1988, pp 458–460, 580.
Robert Gannon, 'The Body Farm', *Popular Science*, September 1997, pp 76–82.

IS MR SMITH HEAVIER THAN MR TAILOR?

When you think of a blacksmith, you think of a big strong person, who is good at large physical movements. When you think of a tailor, you think of a smaller person, who is more skilled in fine dextrous movements. But would you believe that in a few German towns, even today, people called Smith are heavier than people called Tailor? This is not a coincidence. It's all because of how we got our family names.

SO MANY SMITHS 1

'Smith' is the most common name in the English language. In the USA alone, there are almost 2.4 million Smiths — about one per cent of the population.

But the most popular name in the world is 'Zhang'. In China alone, Zhangs make up about 10 per cent of the population, which works out to about 115 million people — which is close to half the population of the USA!

Different societies started using names at different times. According to *The Guinness Book of Records*, the earliest name that we know of dates back to around 3050 BC. This name belonged to an early king of Upper Egypt. In hieroglyphics, it looks like the sign for a scorpion — and if you spoke it, it would probably sound like '*Sekhen*'.

Until relatively recently, in most societies a person would be given just one name. In Christian societies, this name was usually given when they were baptised. This name was called your 'name', or 'forename', or 'given name', or 'Baptismal name' or 'Christian name'.

Early human societies were not as populated as today's cities, so having only a small number of names to pick from wouldn't lead to much confusion. But as populations increased, societies eventually got to the stage where they had too many people with the same name, so the 'family name' was introduced to help identify and differentiate between people with the same 'forenames'.

Your 'family name' is the name that everybody in your family has. It's also called your 'surname' or 'last name'. But it's not always your *last* name. For example, the Chinese have their family name first. So the family name of Mao Tse-Tung was 'Mao'.

SO MANY SMITHS 2

According to the Sydney *Daily Telegraph Mirror* (3 June 1991) one car accident involved too many Smiths.

Pauline Smith was a passenger in a vehicle involved in a roll-over accident. The driver was her daughter, Corolie Smith. She had married Russel Smith, so she kept the same surname. So even though she took her husband's name she was still a Smith. By a coincidence, the witness travelling in another vehicle, was her husband, Russel Smith. The ambulance officer who turned up at the accident was Ross Smith. The attending Police Officer was Constable Trevor Smith. And the name of the tow-truck driver was Robin Smith.

According to the Police Officer, '*the accident could be a bit hard with the paper work*'!

Hungarians also have the family name first, but when they write in the English language, they usually write their family name last.

The earliest family names known were introduced in 2852 BC. In that year, a Chinese emperor ordered that all citizens should adopt family names. In Europe, there was not one single date for the adoption of family names — instead, it was a more gradual process. This process began in 1000 AD, and lasted about six centuries. It started in big cities, and in aristocratic families.

At a certain period of time, it became compulsory to have a family name. In Europe, depending on the country that you lived in, surnames became a legal requirement some time between the 11th and 16th centuries. European Jews living in ghettos got their family names around 1800. Turkey did not make the use of family names compulsory until as recently as 1935!

We don't know why it became compulsory to use surnames and can only speculate. One theory is that the Church and the State wanted to monitor all citizens to make sure that they paid their taxes. (According to some anarchists, the State exists to take money from its citizens — and if they don't pay up, to put them in prison!) As populations increased it became more difficult to trace the many people sharing the same single name.

Some people always have to be different. In the old days, before it became compulsory to have a family name, most people had single names. So the wealthy landowners (some of whom turned themselves into royalty) decided to be different by having a family name.

But now that most people have a family name, the royalty want to appear prominent by using only a single name, such as Queen Elizabeth, or Prince Charles.

A Job by any Other Name . . .

When surnames became compulsory, people had to decide what to call themselves. In many cases, they simply used the names of their jobs. For example, archery was a very important military skill, so people began to call themselves Archer, Bowman, Arrowsmith (a maker of arrows), Fletcher (a maker or dealer in arrows, or bows and arrows) and Bowyer (a maker or dealer in bows).

Of course, a blacksmith would call himself Smith, while a tailor would call himself Tailor. In most cases, the blacksmith's job would require him to be a big, burly fellow, while a tailor could be a thin, rangy chap. The blacksmith would be better at 'gross' motor skills that involved brute strength, while the tailor needed to be better at 'fine' motor skills that involved agility.

Schmids und Schneiders

Günther Bäumler, of the Munich Technical University, claimed that there is a connection between your physique and surname! Bäumler wrote a paper called 'Differences in Physique in men called "Smith" or "Tailor" considered as results of a genetic effect dating back over several centuries'.

In the German language, 'Smith' becomes 'Schmid' or 'Schmied', while 'Tailor' becomes 'Schneider'. Bäumler looked up all the Schmids, Schmieds, and Schneiders, in the telephone books of Munich and Nürnberg. He sent all these 'Smiths' and 'Tailors' questionnaires asking them to rate their strength, height, physique, dexterity, sporting skill, preferred occupation, and so on. Actually, the answer in which he was most interested was their weight, but he didn't let them know this in case they lied. The question asking them about their weight was buried deep in the questionnaire.

Seventy per cent of the people he approached filled out his questionnaire, and returned it. So he ended up with 93 Smiths and each was matched to a Tailor of the same age. When Günther Bäumler did his statistical analysis, he found that the Smiths claimed to be 2.4 kg (5.3 lb) heavier than the Tailors! Furthermore, the Schmidts preferred a line of work that involved physical activity and strength, while the Schneiders claimed that they had chosen professions requiring agility and dexterity. The same trend followed in their choice of hobbies.

Supposedly, say back in the Middle Ages, blacksmiths would have been on average 10 kg (22 lb) heavier than the tailors. But over the many generations, as the blacksmiths' and tailors' descendants bred into the general population, the difference in weight dropped. That's why the difference between Smiths and Tailors today is just 2.4 kg.

Now the science in this study could have been more accurate than it was. There were enough people included for the statistics to be meaningful. However, Günther Bäumler did not actually go out to interview the people involved. They could have lied about their physiques, weight, preferred professions and so on. And even if they had been honest in completing the questionnaires, their bathroom scales might have been inaccurate. It would be interesting if there were more scientifically reliable studies conducted to show that this is a real effect, and not just a one-off effect.

But one question remains — how many of our surnames really do date to that period many centuries ago, when Church and State decided to invade our privacy? And how did we get surnames like Death, Strange, Killer and Flasher!?

REFERENCES

Günther Bäumler (Technische Universität München), 'Difference in Physique in Men Called "Smith" or "Tailor" Considered as Results of a Genetic Effect Dating Back over Several Centuries', *Personality and Individual Differences*, Vol. 1, Pergamon Press, 1980, pp 308–310.

'Excuse Me Smith's the Name', *Daily Telegraph Mirror*, 3 June 1991, p 3.

Encyclopaedia Britannica, CD-ROM, 1996.

Mary Gribbin, 'Tailoring the Facts', *New Scientist*, No. 1246, 26 March 1981, pp 831–832.

SUPER BROCCOLI!

Probably one of the more famous cartoons of all time was penned by Carl Rose, in the *New Yorker* magazine on 8 December 1928. He drew a child refusing to eat his broccoli. The caption underneath written by E.B. White, says '*I say it's spinach, and I say the hell with it*'!

Back in 1928, broccoli had only recently come from Italy into the Santa Clara Valley of Northern California. A lot of Americans didn't like its unfamiliar taste. But even though broccoli has a bitter, sulphurous taste, it's very good for you.

Broccoli Family and Evil Smells

Broccoli is part of the *Brassica* (or mustard) family. Broccoli has a whole bunch of strange-tasting relatives such as brussels sprouts, cabbage, kohlrabi, kale and cauliflower. Many kids don't really like any of them very much — which might be because of their bitter taste and smell. In fact, according to the 1993 *Guinness Book of Records*, the list of the '*most evil of the 17 000 smells so far classified*' has ethyl mercaptan right up near the top. Ethyl mercaptan smells like a mix of burnt toast, sewer gas, garlic, onions and (you guessed it) rotting cabbage.

Broccoli (*Brassica oleracea, Italica variety*) is an annual fast-growing plant. It reaches about 60–90 cm (24–35 in) in height. Broccoli grows well in moderate to cool climates. The slightly bitter immature flower is ready for harvesting in two to five months. It can be eaten raw or cooked. The peeled stalk is sweeter. Try stir-frying it lightly with a little honey in the oil and served with a squeeze of lemon and a dash of tamari.

Bitter Vegies Good For You

It seems the more bitter these vegetables taste, the better they are for your health.

The nutrition scientists have been telling us for years that vegetables and fruit are good for you. One famous study reported in the *British Medical Journal* looked at 11 000 'vegetarians and health-conscious people' for 20 years. It found that people who ate fruit regularly were 24 per cent less likely to die from a heart attack, and 32 per cent less likely to die from a stroke.

WHAT YOU EAT – BRASSICA

Vegetables in the *Brassica* family give you lots of choices about which bit to eat.

In kale, you eat the loose leaves. You can eat the leaves folded into large open heads (cabbage) or into small compact heads (brussels sprouts). In kohlrabi, you eat the stem.

Depending on the particular vegetable, you can eat many different parts. In general, there are five parts to eat. The *root* is edible in carrots. But in celery, the edible part is the *stem*. In spinach, you eat the *leaf*, while in cucumbers, you eat the *fruit*. But in broccoli, you eat the *immature flower bud*.

OTHER GOOD CHEMICALS IN BROCCOLI

Broccoli contains a chemical called *indole carbinol*. Indole carbinol breaks down oestrogen, which is implicated in some breast cancers. According to some nutritionists, a cup of broccoli every second day has enough indole carbinol to protect you.

Strange Chemicals in Broccoli

Broccoli, and all of its bitter-tasting mates, are rich in nutrients — such as proteins, minerals, fibre and various vitamins.

But they also seem to have families of other chemicals that are not nutrients. Instead, these chemicals seem to protect you against various cancers (such as those affecting the colon, breast and lung).

One family of these chemicals is called 'glucosinolates'. Two chemicals in this family that seem to have 'health-giving' effects are 'sinigrin' and 'glucoraphanin'. Sinigrin seems to kill the cancerous cells, while glucoraphanin seems to block cancerous cells. Sinigrin has a bitter taste, while glucoraphanin does not.

BROCCOLI UNPOPULAR WITH KIDS

One theory about why broccoli (and its mates) are not so popular with children might be because of their relatively low sugar levels and bitter taste.

David Laing, a psychobiologist, and his colleagues at the University of Western Sydney, did a study with 600 children, aged between five and 18. They asked the kids to rank eight vegetables, according to how much they liked or disliked them.

Corn was the most popular vegetable, followed closely by peas and carrots. Around the middle of the range were tomatoes, mushrooms and broccoli. However, cauliflower and brussels sprouts were right at the bottom.

Laing thinks that children are less sensitive than adults to the flavours in the foods that they eat. So kids are attracted to highly flavoured junk foods as compensation for the relatively bland sensations that they usually get from food.

This study was part of a larger study to try to work out what actually influences the food preferences of children, and when the taste-sensing apparatus in their mouth and brain becomes mature.

Bitter Anti-Cancer Sinigrin

There's a lot of sinigrin in broccoli and brussels sprouts. The vegetables become too bitter to eat, once the sinigrin level rises above 200 mg/100 grams of sprout.

Ian Johnson, a nutritional physiologist from the Institute of Food Research in Norwich, England, fed sinigrin to laboratory rats. These rats had pre-cancerous cells growing in their gut. Johnson found that just one small dose of sinigrin would actually kill the pre-cancerous cells before they could turn into fully cancerous cells.

Johnson thinks that there is a series of chemical reactions involved. The sinigrin breaks down to another chemical (allyl isothiocyanate, which gives the bitter taste and smell of brussels sprouts). This chemical then triggers pre-cancerous cells to kill themselves, leaving the gut unharmed!

However, we don't know if this chemical pathway happens in humans — it would be a difficult experiment to get past the Ethics Committee.

This chemical is so powerful that it works with just a single dose, not regular doses. So if this chemical works in humans, it's worthwhile to have the occasional brussels sprout.

OFFICIAL US GOVERNMENT HEALTH STATEMENTS

The United States Government has come out with eight official health statements. These deal with the health effects of various foods or food substances. The first five advise you about 'good' things to do:

1. A high calcium intake will reduce your chances of getting osteoporosis.

2. Eating various grain products, and fruits and vegetables that have a lot of fibre, will reduce your risk of getting various cancers.

3. Eating various grain products and fruits which have fibre (especially soluble fibre) will reduce your risk of getting coronary artery heart disease.

4. Eating lots of fruits and vegetables will reduce your risk of getting various cancers.

5. Eating sugar alcohol (such as xylitol) will actually reduce your risk of tooth decay.

The next three points tell what *not* to do:

6. Eating lots of salt will increase your risk of having a high blood pressure.

7. Eating lots of saturated fats and cholesterol will increase your risk of getting coronary heart disease.

8. Having a diet high in fat will increase your risk of getting various cancers.

ENEMIES OF BROCCOLI

When George Bush was President of the USA, he publicly declared his dislike of broccoli. He said, 'I'm rebelling against broccoli, and I refuse to give ground. My mother made me eat it. And I'm President of the US, and I'm not going to eat any more broccoli.'

The broccoli growers of California responded to this threat to their business by sending 10 tons of broccoli to a food bank in Washington DC — but two cartons were specifically sent to the White House.

SUPERTASTERS

Was President George Bush being perverse when he said that he hated broccoli, or was he just telling the truth as he tasted it? Perhaps we can believe him — after all, 25 per cent of Americans are 'supertasters'. A supertaster is a person whose genes make them more sensitive to the bitter chemicals found in some vegetables and fruits.

The story begins in 1931, when Arthur L. Fox of E.I. du Pont de Nemours and Company in Wilmington, Delaware, was making a chemical called PTC (phenylthiocarbamide). By accident, some PTC blew into the air. One of Fox's colleagues inhaled some PTC, and said that it tasted terrible. But Fox tasted nothing at all.

This sparked a lot of interest, and soon the scientists found that we humans could be divided into two groups — tasters and non-tasters. Non-tasters get absolutely no sensation when they taste PTC — they could be chewing on sand, for all the flavour that they get! The scientists soon realised that PTC was slightly toxic, so they switched over to a similar chemical PROP (*6-n-propylthiouracil*).

But a few more decades of research showed that there were two classes of tasters — regular tasters and supertasters. According to Linda N. Bartoshuk, a Taste Researcher at the Yale University School of Medicine, people fall into three groups as far as taste is concerned. In the USA, about 25 per cent of people are supertasters, 50 per cent are regular tasters, and 25 per cent are non-tasters.

Bartoshuk has actually found an anatomical difference between supertasters and regular tasters. On your tongue you have little mushroom-shaped structures called 'fungiform papillae'. The tastebuds live inside these fungiform papillae. Bartoshuk found that supertasters have many more of these fungiform papillae than regular tasters do. You can test this yourself. Wipe some blue food colouring on your tongue. Get a plastic reinforcement ring, one that is usually placed around a standard-size punched hole in paper. Place the ring on your tongue, then look at your tongue in the mirror. If you are not a supertaster, you will see only a few fungiform papillae inside the circle. But if you're a supertaster, there will be more than 25 of them.

This research is definitely not trivial. Supertasters would probably live longer, if they ate these bitter-tasting foods.

Non-Bitter Anti-Cancer Glucosinolate

Gary Williamson, from the same Institute of Food Research, is looking at another glucosinolate called 'glucoraphanin'.

This chemical is found in broccoli and works by a slightly different pathway. When you chew the vegetable in your mouth, the glucoraphanin breaks down into sulphoraphane, which seems to be the active chemical. Sulphoraphane doesn't actually *kill* cancer cells — it just stops them from *growing*. It does this via several different methods.

Sulphoraphane stimulates the liver to make chemicals, which then destroy other chemicals which cause cancer. It also seems to have another effect — it stimulates a part of your immune system called 'Phase II Enzymes'. These Phase II Enzymes also detoxify nasty chemicals and turn them into safe ones.

A recent study at the Dutch Nutrition and Food Research Centre in Zeist looked at volunteers who ate 300 grams (about two-thirds of a pound) of cooked brussels sprouts *every day* for a week (that's a lot of brussels sprouts). The study found that the levels of Phase II Enzymes in the kidney, stomach, liver and gut increased by 50 per cent.

Sprouts Better Than Vegetables

However, most people would find it very difficult (if not impossible) to eat two kg (4.5 lb) of brussels sprouts, or broccoli, each week. Stay cool, there is an easy way out — eat sprouts. Remember that glucoraphanin turns into sulphoraphane. The sprouts (growing seeds) of broccoli and brussels sprouts contain up to 50 times more glucoraphanin than the mature vegetable.

Different cultivars of broccoli have different levels of glucoraphanin. One cultivar, Green Comet, has less than 10 micromoles/gram of glucoraphanin. But other cultivars of broccoli (such as Mercedes and Emperor), have about 60 or 70 micromoles/gram. Trixie has the highest level ever measured — 150 micromoles/gram.

Why is Something so Bitter so Good?

The vegetables in these families probably evolved these bitter chemicals to discourage animals from eating them. So how did these chemicals come to be good for humans? As we have spent hundreds of thousands of years eating a wide variety of fruits and vegetables it's possible that over that time, we have adapted to, and actually *need* to, regularly eat plant toxins.

Here's the irony. Supermarkets and food chains are now getting the growers to breed these bitter *Brassica*-family plants to have *milder* flavours — flavours that are more nutty, than bitter. But as the flavours get milder, so the levels of protective chemicals drop.

Vegetable or Tablet

Of course, some enterprising person will simply try to remove the bitter-tasting-but-good chemicals, and sell them to you in little capsules. But if the last few hundred years of nutrition have taught us anything, it's that you need to eat the entire vegetable to get all of the good health effects.

Maybe, once we know more about these 'health-giving' chemicals, the genetic engineers can modify plants to have more of these 'good' chemicals.

Probably the worst thing about this latest piece of health advice, is that once again, it proves that your parents were right.

HISTORY OF BROCCOLI AND ITS FRIENDS

The Ancient Romans regularly ate broccoli. In 1533, Catherine de' Medici introduced many vegetables into France, including broccoli, savoy cabbage, and globe artichokes. It took another two centuries until 1721, before broccoli made its way into England. Soon after, broccoli emigrated to America.

It seems as though brussels sprouts were probably grown in Belgium around the 1200s. But the first recorded description of brussels sprouts was written in 1554 by the Dutch botanist, Rembert Dodoens.

Kohlrabi was first described in Europe in the 1500s.

WORLD RECORD BRASSICA

According to the 1993 *Guinness Book of Records*, the biggest cabbage on record was grown in 1989 by B. Laverly of Llanharry in Great Britain. It weighed a massive 56 kg.

BROCCOLI VS INSECTS

How did the *Brassica* family evolve to the current state where they are loaded with bitter chemicals?

One popular theory says that these chemicals taste terrible to insects and other predators. But some insects have evolved special detoxifying systems to neutralise these chemicals. In fact, some insects find these bitter *Brassica* chemicals very attractive.

But the *Brassica* plants had an extra trick. Their glucosinolates are very attractive to a few parasitic wasps — who come and eat the other insects which are bothering the *Brassica* plants.

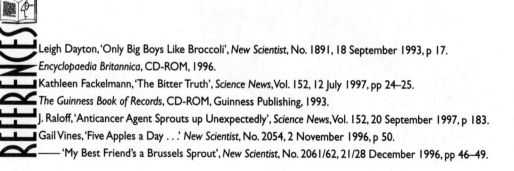

REFERENCES

Leigh Dayton, 'Only Big Boys Like Broccoli', *New Scientist*, No. 1891, 18 September 1993, p 17.

Encyclopaedia Britannica, CD-ROM, 1996.

Kathleen Fackelmann, 'The Bitter Truth', *Science News*, Vol. 152, 12 July 1997, pp 24–25.

The Guinness Book of Records, CD-ROM, Guinness Publishing, 1993.

J. Raloff, 'Anticancer Agent Sprouts up Unexpectedly', *Science News*, Vol. 152, 20 September 1997, p 183.

Gail Vines, 'Five Apples a Day . . .' *New Scientist*, No. 2054, 2 November 1996, p 50.

—— 'My Best Friend's a Brussels Sprout', *New Scientist*, No. 2061/62, 21/28 December 1996, pp 46–49.

LIFE ON TITAN

The only place that we've ever found life is down here on planet Earth, but some recent photos of Titan, one of the moons of the ringed-planet Saturn, hint that we might find life there. Titan is the only known moon that has a thick atmosphere, and the chemicals in that atmosphere are the chemicals of life.

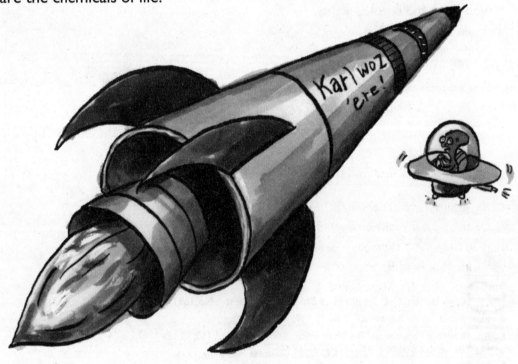

Our Solar System

There are nine planets in our solar system. Closest to the Sun are the four rocky planets (Mercury, Venus, Earth and Mars). The rocky planets are all fairly small and have two or fewer moons. Then, further out from the Sun, is another bunch of four planets. These are the four gas giants (Jupiter, Saturn, Uranus and Neptune). They each have a set of rings (although Saturn has the most impressive), and they each have a set of moons orbiting them. In fact, each of the outer gas giants is like a mini solar system. Finally, furthest out from the Sun is another small rocky planet, Pluto. (Actually, some astronomers don't count Pluto as a planet, but as a giant captured asteroid.)

There are about 60 or so moons altogether in our solar system, but only one of them has an atmosphere similar to our own. It's Titan, which is very big. In fact it's bigger than the planet Mercury, bigger than our Moon, a bit smaller than the planet Mars, and about two-and-a-half times smaller than the Earth. If it wasn't a moon of Saturn, it could easily be a planet in its own right.

Titan's Atmosphere

Titan's atmosphere is about 60 per cent denser than Earth's atmosphere. No other moon in the solar system has an atmosphere even one-thousandth as dense as our atmosphere. Earth's atmosphere is mostly nitrogen (80 per cent), with oxygen and a few other gases thrown in. Titan's atmosphere, like Earth's, is mostly nitrogen (82–95 per cent). Methane (2–10 per cent) is the second most abundant gas. The other gases include ethane, acetylene, water vapour

GRAVITATIONAL SLINGSHOT = GRAVITY-ASSIST MANOEUVRE

The Titan/Centaur rocket that launched *Cassini/Huygens* did not have enough grunt to send it directly to Saturn. So the rocket has to do a complicated series of gravity-assist manoeuvres. In each one, it steals a small amount of energy from a planet, and uses this energy to speed itself up.

The dates when it swings past planets are:

Venus, 21 April 1998 (gaining 7 km/sec).

Venus, 20 June 1999 (gaining 5.7 km/sec).

Earth, 16 August 1999.

Jupiter, 30 December 2000.

It should arrive at Saturn on 1 July 2004.

The environmental effects of this stolen energy are vanishingly small. When the *Voyager* spacecraft flew by Jupiter, it sped up by 16 km (10 miles) per second – and slowed down Jupiter by only 30 cm (12 inches) every trillion years!

and various hydrocarbons. The atmosphere is about 10 times deeper than ours. Titan is still one of the most mysterious objects in our solar system, even though we have been looking at it for a long time.

VOYAGER 1 MAKES A SACRIFICE

The *Voyager 1* and 2 spacecraft were launched in 1977. This was the perfect time for them to use the gravity-assist manoeuvre to visit most of the outer planets in just 12 years, not 50! The outer planets are lined up in such an ideal alignment only very rarely — once every 175 years. *Voyager 1* visited Jupiter, and Saturn, and then headed up out of the solar system. *Voyager 2* continued to leapfrog (via the gravity-assist manoeuvre) to the outer gas giants, Uranus and Neptune. *Voyager 1* deliberately gave up this capability just to get a good look at Titan. This sacrifice was definitely worth it. The *Voyager 1* data told us that we should have a close look at Titan.

Titan's Discovery and Exploration

Titan was first seen and named by the astronomer, Christiaan Huygens, back in March 1655. He saw a largish circular blob that went around Saturn. In 1908, the Catalan astronomer José Comas Solá saw through his telescope that Titan was darker at its edges than at its centre. He thought that this meant that Titan had an atmosphere. In 1944, the American astronomer, Gerard P. Kuiper analysed the sunlight reflected from Titan and he found methane.

In 1980, the *Voyager 1* spacecraft came within 4000 kilometres (2485 miles) of Titan. *Voyager 1* told us that Titan had a thick atmosphere (1.6 times denser than our own), and that like Earth's atmosphere, it had several layers. It also analysed the gases in the upper atmosphere. But the atmosphere was too thick to see through, so we had no idea of what kind of surface lay

underneath that thick photochemical smog that is the atmosphere of Titan.

The atmosphere seemed to be very similar to the gases that were in our atmosphere in the early days of planet Earth. So when the planetary scientists proposed to send another spacecraft to Saturn, it seemed a good idea to get a close look at Titan at the same time. They started planning this mission in 1982. There were tantalising hints about the similarities between Earth and Titan — but there were also huge gaps in our knowledge. The planetary scientists wanted to know more.

Radar and Infra-Red See Oceans?

In 1989, astronomers used the NASA Deep Space Network dish at Goldstone, California to bounce radar signals off the surface of Titan. The radar signals that they received were not very detailed, because Titan is so far away. But the radar pictures revealed one great surprise. They showed

THE TITANS

In mythology, the Titans were enormous and very strong creatures who were the children of the gods, Uranus and Gaea. They were also called the 'Elder Gods', and they ruled the universe for a long time after they had overthrown their father, Uranus. They had a very complicated family life, especially when it came to having relationships, and to making children.

To make it even more complicated, the name Titan is used for the 12 children of Uranus and Gaea, and also for some of their grandchildren (eg, Prometheus, Atlas, Helios and Hecate).

There were 12 Titans — Cronus (who castrated and deposed his own father, Uranus, and then led his fellow Titans against his own son, Zeus, and his fellow Olympians, and lost, and was then deposed), Rhea (who married Cronus, and then gave birth to the gods of Olympus, some of whom deposed the Titans), Oceanus (linked to the stream surrounding the world), Coeus (or Coius), Lapetus (the father of Atlas, who held up the heavens, and Prometheus, who stole fire and gave it to the humans), Hyperion (a sun god, and the father of the Sun, the Moon and the Dawn), Creus, Theia (or Thea, who was linked to the sky), Mnemosyne (the goddess of memory, who also gave birth to some of the children of Zeus, these children being the nine Muses), Phoebe, Tethys (linked to water, and wife of Oceanus), and Themis (an Earth goddess, and the goddess of divine justice, and one of the wives of Zeus, and who gave birth to the Hours and the Fates).

After they had deposed Uranus, the Titans ruled the universe. But then Zeus (who was the son of a Titan) and his fellow Olympians defeated the Titans after 10 years of bitter warfare. This war is called the Titanomachia (or Titanomachy). Most of the Titans were sent to the underworld. But a few (Atlas and Prometheus) who sided with Zeus and fought against Cronos, were allowed to stay on Earth.

that the surface of Titan was not the same everywhere, but was made from different materials — perhaps land and seas.

In October 1994, astronomers used the recently refurbished Hubble Space Telescope to look at Titan using near infra-red light, rather than visible light. It's easier for near infra-red light to penetrate Titan's atmosphere, and harder for visible light. These pictures could not have been taken with a ground-based telescope because our atmosphere would absorb the near infra-red radiation. So the Hubble Telescope, high above the atmosphere, was in the perfect location. The resolution (or sharpness) of the pictures wasn't very high, but they showed more detail than the 1989 radar pictures. The pictures showed something that could be a continent the size of Australia, and other areas that could be seas.

TITAN FACT FILE

Titan is the second largest moon in our solar system (after Ganymede of Jupiter). The atmospheric pressure is 60 per cent higher than ours — roughly equal to the pressure at the bottom of a backyard swimming pool. It is 5150 kilometres (3200 miles) in diameter — about 40 per cent of the Earth's size. It orbits Saturn at a distance of 1 221 830 kilometres (759 210 miles), and takes 16 Earth days (actually 15.95 days) for one complete orbit and one complete rotation. Like Europa and our own Moon, it keeps the same face to its mother planet. Titan is not very dense — only 1.88 times denser than water. This means that Titan probably does not have much of a rocky core.

We still don't know very much, but we know that Titan is very cold — about 180°C below zero. At that temperature, water is a solid lump. But around that temperature, methane and ethane can exist as solid, liquid or gas. So on Titan, you could have methane rain or methane snow, seas of methane and ethane, or perhaps even brightly coloured glaciers of solid hydrocarbons. You have probably heard of the Greenhouse Effect here on Earth. Titan is warmer than it would otherwise be, thanks to a Greenhouse Effect that is powered by the light of the distant and weak Sun. Titan's orange-coloured atmosphere keeps it a not-so-cosy minus 180°C.

Something Weird on Titan

So while we know a bit about Titan, there is still a lot that we don't know. According to the chemists, the methane should have all turned into ethane, hydrogen cyanide, and other organic chemicals by now — and yet, there is still methane on Titan. So some process we don't understand is making methane in the atmosphere of Titan. That unknown process could be bacteria eating, excreting, living and dying.

HUYGENS AND CASSINI

Christiaan Huygens lived from 1629 to 1695. He was a Dutch astronomer who discovered both the rings of Saturn, and in 1655, Titan. He also invented the pendulum clock, and came up with the wave theory of light. Jean-Dominique Cassini lived from 1625 to 1712. He was a French–Italian astronomer, who first found a gap between the rings of Saturn. (Today, we have found about 100 000 gaps.) It was called the Cassini Division in his honour.

Ingredients of Life

Life as we know it, needs three basic ingredients — raw materials, three states of matter, and energy. First, you need the raw materials, which could include nitrogen, carbon or hydrogen or oxygen. (Someday we might discover a form of life that doesn't need matter and that runs on 'pure energy', but at the moment, raw matter is essential.) Second, you need the three states of matter, such as solid, liquid and gas. On our planet, water can exist in these three states. As it shifts from one state to another, it can carry other chemicals with it — and so you get a complete life cycle. Finally, you need a source of energy, like the Sun, to pump the whole cycle along.

Life Cycle

To get an idea of a 'Life Cycle', take a look at an atom of carbon. It might start off in a molecule of carbon dioxide floating in the air. It might then be breathed in by a blade of grass, and then become part of a carbohydrate molecule in the grass. Our carbon atom would go from gas to a liquid state, before ending up in a solid state. The energy for this process would come from the Sun. A cow might eat that solid carbohydrate in the grass, and turn it into milk — and so our carbon atom might be part of a lactose molecule in the cow's milk. I might drink that milk, metabolise it, and then breathe out our carbon atom as part of the gas carbon dioxide from my mouth. The cycle would continue as the carbon dioxide is taken up by a corn plant in my garden. So it's very handy to have solid, liquid and gas on your planet or moon, if you want life. Our Moon has no life (as far as we know). It has two of the three requirements — raw materials, and energy from the Sun. But it has only one of the three states of matter. The raw materials are all solid, and tied up as rock or dust — and there's no liquid or gas to let a life cycle happen.

On Titan, we definitely have two of the three ingredients for life — and maybe the third. First, there are the right raw materials — hydrocarbons of many types. Second, it also seems as though we have the right states of matter — solid, liquid and gas. But what about the energy? That is the problem. After all, at that distance out, the Sun is just a very bright star. Saturn is about 10 times further away from the Sun than the Earth, so the solar energy is about 100 times weaker, or only 1 per cent of its level here on Earth. High noon is probably as bright as a moonlit night here on Earth. It is possible that chemical reactions occur on Titan, but 100 times more slowly. Or maybe bolts of lightning provide the energy.

Spacecraft Flies Out

We're hoping that a spacecraft now on its way to Saturn and Titan might tell us much more about Titan. (I, and a million other people, have my name on this spacecraft.) On 15 October 1997, the *Cassini/Huygens* spacecraft was successfully launched on a 6.7-year journey to Saturn (on top of a Titan IVB/Centaur rocket). It will spend most of its time asleep, but, on average, the 'health' of the spacecraft will be checked every six months. *Cassini/Huygens* is one of the heaviest spacecraft ever launched to another planet — 5.820 tonnes (5.728 tons). Over half of that mass is fuel. It's big as well — 6.7 metres (22 feet) long.

Cassini/Huygens is the result of a co-operation between NASA and the European Space Agency.

NASA built the *Cassini* Orbiter mother ship. It will arrive at Saturn on 1 July 2004. It will slow down (by burning a lot of fuel), so that it can be captured by the gravity of Saturn. On 6 November 2004, the *Cassini* Orbiter will shed the *Huygens* Probe. *Cassini* will then spend from 2004 to 2008 orbiting Saturn some 60 times. It should

MY NAME AT SATURN

Before the launch of *Cassini/Huygens*, NASA made a great offer to the citizens of the world. Just send them your signature, and they would scan it onto a CD-ROM, which they would put on the *Cassini* part of the spacecraft. There would be one million entries in all. I sent off a whole bunch of signatures (our family, and all the other kids in my kids' class) to NASA — and now they're heading for Saturn.

NOT JUST PEOPLE'S NAMES AT SATURN

Not all the words on the CD-ROM are just people's names. Florence Dugoua, aged 30, wrote, 'Tall female French Earthling seeks tall handsome alien, romantic if possible'. A certain 'President Bill Clinton' invited (in fluent French!) Titan to 'become the next state of the United States'. Cristelle Levrat's note was short: 'Hello little green worms.'

do some 33 fly-pasts of Titan — some as close as 850 kilometres (528 miles). The European part is the *Huygens* Probe that will be dropped onto Titan. It weighs about 319 kg (703 lb). Only 44 kg (97 lb) of this is actual scientific payload — the rest is heat shields, parachutes, etc. At the end of a 21-day cruise after separating from the *Cassini* mother ship, *Huygens* will slam into the atmosphere of Titan at around 20 000 km/hour (12 400 miles/hour). This should happen on 27 November 2004. It will slow down thanks to its heat shield and three parachutes, and then spend 2.5 hours dropping down to the surface of Titan.

Future Discoveries

There is so much that we want to know about Titan. We want to know if there are seas of liquid ethane, and if there is a rain of methane. Where does the energy come from to drive the chemistry that happens on Titan, and how bright is the Sun at the surface? Is there a solid part of the surface of Titan, and does it have rivers, mountains and valleys? Exactly what gases are in the atmosphere of Titan, and in what ratios, and how does this vary with altitude? How do the winds blow — at the surface, and higher up? What are the clouds made of?

Huygens has six separate science experiments using equipment which includes thermometers and barometers, chemical analysers and gas chromatographs, magnetometers, altimeters, cameras and even a microphone. If all goes well, these experiments will answer most of these questions.

But they probably won't answer the big question — could life possibly exist on Titan? That will be for another spacecraft with different experiments. Perhaps then we might be lucky enough to find the first evidence of life outside our planet. It might only be bacteria, but even alien bacteria would be amazing.

PRIEST SAVES BOOK FOR *CASSINI*

A recipe found in an old book in East Germany has helped build an essential part of the *Cassini* Orbiter.

This essential part is a spectrometer, which was being built by Hans Lauche of the Max Planck Institute for Aeronomy in Katlenburg-Lindau, in Germany. The spectrometer will measure the ratios of various isotopes of hydrogen in the atmospheres of the ringed planet, and of its giant moon, Titan. This information will tell NASA scientists much about the relative evolution and growth of Saturn and Titan. The spectrometer will measure light, using an amplifier made of a composite bond of glass and a ceramic.

This composite bond was a problem. Modern western ceramics, such as oxides of beryllium or aluminium, hardly expand or contract at all, in response to temperature changes. However, glass expands a lot when the temperature increases. So a composite of glass and a modern ceramic would simply split at the join, as the temperature changed. Lauche simply could not find, anywhere in the western world, an old-fashioned ceramic with similar temperature/expansion properties to glass.

However, Eastern European countries used these ceramics as recently as 10 years ago. Unfortunately, when Germany was unified in 1990, many of the libraries in East Germany simply threw out old books, in the belief that the new books they would shortly receive would somehow be better. Martin Weskott, a Lutheran priest in Katlenburg, is also known as the 'Book Pastor'. Over the last decade, he has personally saved some 700 000 old books from destruction, and stored them in an old monastery. Lauche was desperate for information about the old ceramics, and he knew that if anybody had a book with the needed knowledge, Weskott would. The guess paid off. Weskott did have a book with the required information, and this knowledge was used to build the spectrometer now on *Cassini*.

Ken Croswell, 'A Moon With Atmosphere', *New Scientist*, No. 1788, 28 September 1991, pp 33–36.

Michael A. Dornheim, 'Cassini Mission to Saturn Caps Era of Grand Spacecraft', *Aviation Week and Space Technology*, 9 December 1996, pp 71–80.

Jeff Hecht, 'Radar Echoes Reveal Ice Continents on Titan', *New Scientist*, No 1674, 22 July 1989, p 16.

Ralph Lorenz, 'Lifting Titan's Veil', *New Scientist*, No. 2077, 12 April 1997, pp 34–37.

Debora MacKenzie, 'Priest Supplies Saturn Probe With Missing Ingredient', *New Scientist*, No. 2060, 14 December 1996, p 7.

Tobias Owen, 'Titan', *Scientific American*, February 1982, pp 76–85.

Josh Rogan, 'Bound for the Ringed Planet', *Astronomy*, November 1997, pp 36–47.

Carl Sagan and Stanley F. Dermott, 'Tidal Effects of Disconnected Hydrocarbon Seas on Titan', *Nature*, Vol. 374, 16 March 1995, pp 238–240.

http://www.estec.esa.nl/spdwww/huygens/html/newmain.html — April 1998

LIFE ON EUROPA

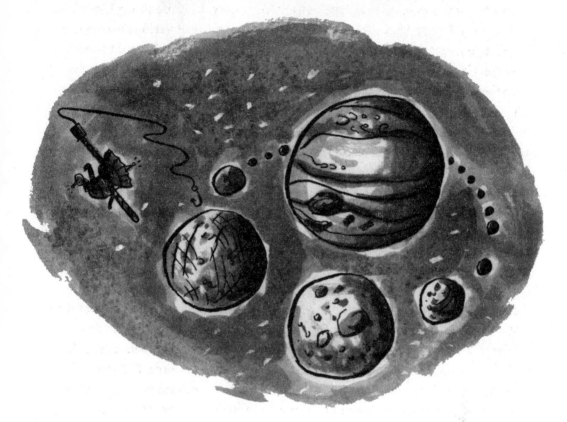

Oceans are wonderful places. In fact, according to John R. Delaney, an oceanographer at the University of Washington: 'An ocean is the womb of a planet.' An ocean is where life starts. The last time we discovered a major ocean was about 500 years ago, when Balboa supposedly discovered the Pacific Ocean. But our little blue planet, Earth, might not be the only place with oceans. There might be another ocean, 800 million kilometres away, on the smoothest object in the solar system.

Scientists running the *Galileo* spacecraft, which is currently orbiting around Jupiter, think that they might have found an ocean of water on one of the moons of Jupiter. An ocean of water might mean life as we know it (this life would probably be bacteria, rather than kangaroos or elephants!).

Galileo and His Telescope

On 10 January 1610, Galileo Galilei took his recently built telescope to his favourite window seat in his house in Florence, and aimed it at the planet Jupiter. His little telescope could 'make' objects appear 1000 times bigger. To his astonishment, he discovered three little white blobs right next to the planet. Over the next month he discovered that these blobs seemed to move, and he found a fourth blob. He had discovered four of the 16 currently known moons of Jupiter.

Jupiter is the biggest planet in the solar system. By itself, it accounts for 90 per cent of the mass of all the planets and moons in our solar system. Jupiter has a gravitational field about 300 times stronger than that of the planet Earth.

The Moons of Jupiter

The innermost moon of Jupiter, Io, has a slightly elliptical (or egg-shaped) orbit around Jupiter. The combination of Jupiter's enormous gravity, and this elliptical orbit, means that during each orbit, the surface of Io actually pumps up and down by about 50 metres. This enormous flexing generates heat, called Tidal Heating. On Io, this heat is so great that the moon is continually erupting with volcanoes of molten sulphur.

The second moon out from Jupiter, Europa, also has tidal heating, but to a lesser degree. It seems to have volcanoes of either icy slush or water. Europa, which takes 3.6 days to orbit Jupiter, is the moon that might have oceans of liquid water — thanks to this tidal heating.

EUROPA FACT FILE

Europa is about 3138 kilometres (1950 miles) across, and orbits Jupiter at a distance of about 670 900 kilometres (416 877 miles). It keeps the same side facing Jupiter, and so its day (rotational period) is as long as its Europan year (time taken for a complete orbit) — about 3.55 Earth days. It has a relative density of 3.01, so there is probably a large rocky core.

Europa is named after a Phoenician princess who was kidnapped by Zeus, the King of the Gods, then taken to Crete. The moon Europa is just a little smaller than our own Moon. It has an incredibly tenuous atmosphere of oxygen — about 100 billion times thinner than our own atmosphere.

HOW TO HEAT A PLANET

Planetary scientists hadn't thought a lot about tidal heating, until the *Voyager* photos of Io arrived back on Earth. They did know of two other ways to warm a planet. First, they knew of kinetic heating. When a fast-moving rock, comet or asteroid slams into a planet, much of that kinetic energy turns into heat. Second, they knew of the heat given off when radioactive elements (such as uranium, potassium and thorium) decay. But they were stumped when they saw the photos of Io, with spurting volcanoes of sulphur. They soon realised that tidal heating was another way to heat up planets.

Voyager Visits Europa

We didn't know much about Europa until 1979, when the *Voyager 1* and *Voyager 2* spacecrafts zipped past Jupiter. They didn't get really close to Europa — *Voyager 1* sailed past some 734 000 kilometres away, while *Voyager 2* flew past at a range of 206 000 kilometres. From these great distances, the *Voyager* photos don't show much detail. But these early photos do show that the surface of Europa seems to be made entirely from cracked ice. Europa looks like a strangely fractured eggshell, or a cracked billiard ball. Its surface is covered with strange lines, up to 3000 kilometres long, and up to 70 kilometres wide. Most of these strange lines were straight, but some of them were jagged or curved, and they all cross each other every which way.

Galileo Visits Europa

Our next big jump in knowledge about Europa occurred after 7 December 1995, when the *Galileo* spacecraft arrived at Jupiter after a six-year trip. (*Galileo* was the first spacecraft to orbit one of the outer planets of our solar system.) Its job was to loop around Jupiter repeatedly, and to

GEM

Galileo's mission was supposed to come to an end in December 1997. But it was extended another two years until December 1999, because of the amazing discoveries that had been made. The GEM (or *Galileo* Europa Mission), includes another eight close fly-pasts of Europa.

Then *Galileo* will make four fly-pasts of the moon Callisto, and then head for almost certain suicide. It will repeatedly fly through the belt of charged particles that make a 'doughnut' around Jupiter, in the orbit of Io. These highly charged particles are expected to destroy the electronics of *Galileo*, but not before *Galileo* sends back the closest pictures of Io ever taken.

swing close to most of the moons to have a really good look. On one loop, it flew only a few hundred kilometres above the surface of Europa — about 1000 times closer than the *Voyager 2* spacecraft. While *Voyager 2* could see details as small as 4 kilometres across, *Galileo* could see something the size of a small truck.

The rocket that launched *Galileo* didn't have enough power to send the spacecraft directly out to Jupiter. So *Galileo* looped around Venus (twice) and Earth (once) in a gravitational-slingshot manoeuvre, to pick up extra speed. (See *Life on Titan* for more on this manoeuvre.) When *Galileo* flew past Earth, it flew over Australia and the Antarctic. It took photographs of the ice in the Antarctic.

Ice on Europa

The surface of Europa looks just like this ice in the Antarctic! The surface of Europa seems to be a thin ice crust, covering either liquid water or frozen slush.

This discovery of lots of frozen water elsewhere in our solar system is amazing. This ice is genuine water ice, not some obscure hydrocarbon ice.

The *Voyager* spacecraft took pictures of Europa from hundreds of thousands of kilometres away, and showed us mysterious lines. But the *Galileo* pictures from a few hundred kilometres away, showed us that the lines are ridges of ice.

Some parts of Europa look like jumbled fields of icebergs. In some places, there are giant icebergs, 8 kilometres across, that have been pulled apart and rotated. These rafts of ice look as though they've been broken off from bigger lumps of ice, and then tilted or spun, and then melted back

into position. Some of the material on the surface looks as if it's been spewed out of an ice volcano. There are big slabs of ice that seem to have been partially melted, and then frozen. And everywhere there is the amazingly complex system of criss-crossing ridges, as though the ice has been repeatedly pulled apart, or pushed together.

Some scientists think that the size of the ridges means that Europa has a brittle outer layer of ice, only 200 metres thick. Under that could be a layer of warmer ice, and underneath that, liquid water. But we can't yet be sure.

Water Ice, Yes, Liquid Water, Maybe

The scientists do agree on two facts. First, they agree that the surface ice is water ice. Second, they agree that there was definitely liquid water on Europa in the past. But is there liquid water underneath the ice today? Nobody knows, and *Galileo's* instruments are not the right ones to tell us.

There are two pieces of evidence that suggest there might still be liquid water underneath the ice. First, in some areas there are virtually no craters on the surface of Europa. Jupiter has an enormous gravitational field, and like a cosmic vacuum cleaner, sucks in comets and other small heavenly bodies like crazy. The other moons of Jupiter have pockmarks all over their surface from millions of comet impacts. But some parts of the surface of Europa have hardly any craters.

Second, Europa is the smoothest body in the solar system. The highest mountain on Europa is less than a kilometre high. Anything higher seems to 'collapse' under its own weight.

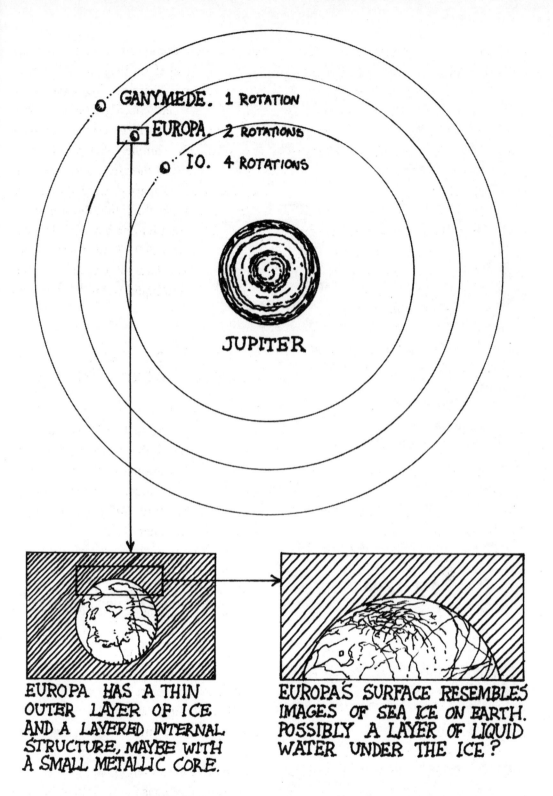

GANYMEDE. 1 ROTATION

EUROPA. 2 ROTATIONS

IO. 4 ROTATIONS

JUPITER

EUROPA HAS A THIN OUTER LAYER OF ICE AND A LAYERED INTERNAL STRUCTURE, MAYBE WITH A SMALL METALLIC CORE.

EUROPA'S SURFACE RESEMBLES IMAGES OF SEA ICE ON EARTH. POSSIBLY A LAYER OF LIQUID WATER UNDER THE ICE?

Perhaps the surface is slightly plastic and flows easily. Oceans of liquid water just under the ice is one obvious explanation — the core of a planet or moon is always hotter than the surface, and if there is ice on the surface, there could be liquid water underneath.

COSMIC DANCE OF THE MOONS

The three inner moons of Jupiter are locked in a strange cosmic gravitational dance.

To get a feel for this dance, look at our Moon. As the Moon goes around the Earth, it drags the water with it, which gives us the tides. These tides create friction on the floor of the ocean, and so as time goes by, the Earth spins a little more slowly. This slower spin means that Earth loses a little energy. This energy is transferred from the Earth to the Moon, and so the Moon moves a little further away. The Moon is moving away from the Earth at about 4 metres per century.

Jupiter and its three inner moons have a similar cosmic dance, but one which is far more elegant and ornate. In the beginning of the Solar System, soon after Jupiter and its moons popped into existence, all these moons moved independently of each other. Over a few hundred million years, Jupiter transferred some of its energy to Io, which began to slowly spiral outwards (just like the Moon still moves away from the Earth).

Io continued to move outwards from Jupiter until it ran into the gravitational pull of the next moon, Europa. The gravitational pull of Europa is tiny — less than one 10 000-thousandth of the pull of Jupiter. But it was enough to lock the orbits of the two moons together.

Every time Europa makes one complete loop around Jupiter, Io makes two. Every 1.8 days, Io comes just 'underneath' Europa (that is between Europa and Jupiter). Suppose that Io is either a little ahead, or a little behind, instead of directly 'underneath' Europa. Either way, Europa gives Io a little pull, which keeps their orbits in that two-to-one relationship.

The mathematicians tell us that this pull also shoves both Io and Europa into more elliptical orbits. This means they get more tidal heating, as their crusts move up and down.

Io and Europa were now locked together as a pair, and they were still robbing rotational energy from Jupiter. So, as a pair, they kept on spiralling out, moving slowly away from Jupiter.

After about a billion years, they were far enough away from Jupiter to be 'pulled' by Jupiter's third moon, Ganymede. Soon Ganymede joined the gravitational dance. So now when Ganymede goes around Jupiter once, Europa goes twice and Io goes around four times.

In another ten billion years or so, the three moons will have moved out enough to get Callisto, the fourth moon of Jupiter, to join in this cosmic dance.

Europa Observer

The *Galileo* findings were so exciting that NASA has announced that in 2003 it will launch a spacecraft called the *Europa Observer*. It will have a sophisticated long-wavelength radar system that can look right through the ice, and find any liquid water underneath — if there is any. It will also probably have a laser instrument to look at the surface of Europa, to measure exactly how much the surface flexes up-and-down each day. And of course, it will have a few cameras (both visible light, and infra-red light). Back in 1977, when the *Voyager* spacecraft were launched, strange new life forms were discovered on our ocean floor, a few kilometres down. There were enormous white clams and giant three-metre-long worms without a mouth or an anus. These creatures were the first discovered that did not rely on photosynthesis to exist. They did not need the energy of the Sun. Similar kinds of creatures could live off the tidal heating in the oceans of Europa. Arthur C. Clarke wrote *2001: A Space Odyssey*. In the follow-up, *2010*, a mysterious message arrives at Earth and warns us, 'All these worlds are yours, except Europa. Attempt no landings there'. But Clarke has since said that we have to explore this world that may have water on it. If there's liquid water, there is almost certainly life there — and that would be really exciting.

THE 'SCIENCE' OF RIDGES, OR 'RIDGOLOGY'

As Europa spins around Jupiter every 3.6 days, the surface flexes (or pumps up-and-down), thanks to the enormous gravity of Jupiter. These tides are huge. One theory for Europa's ridges is that because ice is fairly brittle, it can crack. If there is water underneath, it would seep up and fill the crack, and then partially freeze. And then when the next tide squeezes the crack shut, the slightly-frozen mush would be pushed out again, giving us the ridges that the *Galileo* photos show us.

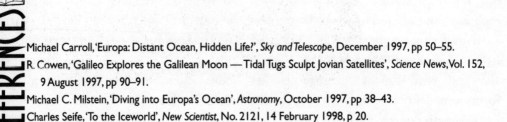

Michael Carroll, 'Europa: Distant Ocean, Hidden Life?', *Sky and Telescope*, December 1997, pp 50–55.
R. Cowen, 'Galileo Explores the Galilean Moon — Tidal Tugs Sculpt Jovian Satellites', *Science News*, Vol. 152, 9 August 1997, pp 90–91.
Michael C. Milstein, 'Diving into Europa's Ocean', *Astronomy*, October 1997, pp 38–43.
Charles Seife, 'To the Iceworld', *New Scientist*, No. 2121, 14 February 1998, p 20.
http://www.jpl.nasa.gov/galileo

FLATTENED FAUNA

U nfortunately, the way that most city people are introduced to our natural wildlife is by seeing them dead on the road. These dead creatures are called 'roadkills'. When animals get run over by a vehicle, not only do they usually lose their life, they usually lose their third dimension as well — which is why they're also called 'flattened fauna'.

Koalas Die

In the holidays when we head out onto the roads, we slaughter many of our native animals. The combination of a wide highway and a slow-moving inexperienced young animal is often fatal. Most of our Australian animals are nocturnal, and are active between dawn and dusk. They're mostly grey or brown-grey, so they're hard to see. Luckily the vehicle traffic does drop off between midnight and six am, but even so, many of our animals die on the roads.

According to the Australian Koala Foundation, a tragic, but typical, road slaughter happened at Iluka, east of Grafton, NSW. The Iluka Peninsula used to have many koalas, but then the road running through the middle of the peninsula was upgraded. Over the next year, one third of all the koalas in the area died — thanks to the increased road traffic.

Flattened Fauna – The Early Days

Cars started appearing on the roads around 1910. Animals had no history of how to treat these strange new moving objects. Back in 1933, one driver wrote that on an 80-kilometre (50-mile) stretch of road near Boise, Idaho, he counted 598 dead rabbits. However, not many people realised the scientific importance of roadkill.

But today, research into this previously unrecognised field of 'flattened fauna' is very active, and many books and scientific papers have been written on it.

Knutson – A Pioneer

One of the early writers about flattened fauna was Roger Knutson, a professor of biology at Luther College in Decorah, Iowa. Back in the early 80s, he wrote a book called *Common Animals of Roads, Streets, and Highways: A Field Guide To Flattened Fauna.*

He would often discuss roadkills in his lectures. He soon realised that a 'field guide' to roadkills would be really handy. So, one of the handy features of his book is the set of questions which helps you identify just what sort of animal it was before it got flattened. You just work your way through the questions. For instance, if you answer 'no' to the question *'Fur or feathers?',* but 'yes' to the question *'Is the blob more than twelve times longer than it is wide?'* — then you've got a dead snake.

Knutson also introduced the word 'enroadment' into the English language, to describe the process of being squashed, and then dried in the sun.

These roadkills happen for a couple of reasons. First, the animal usually has a specific defence pattern which doesn't work very well with cars and trucks. Secondly, a human runs the poor animal down.

In different areas, different animals have different defence patterns. For example, in Iowa, the most common roadkill is the striped skunk. Its defence pattern is to just stay there and emit its famous odour. Not very effective against a speeding family car!

But in Texas, the armadillo is number one in the flattened fauna scene. Its defence is either to roll up into a little armoured ball, or to jump vertically into the air to make itself look bigger. This defence is great against cats and dogs, but against a 20-tonne, 18-wheeler semitrailer, it has all the structural integrity of a flowerpot.

Many snakes end up as roadkills, because at the end of a long hard day of slithering, they like to stretch out on a warm flat wide rock — like a road.

Knutson's book also has a section of 'before' and 'after' photographs. For example, when the book deals with the 'vagabond painted turtle', the sad caption next to the 'after' photograph describes the turtle as looking 'much like a dull-coloured pile of crushed crockery'. The pathetic flattened turtle is mostly circular, with just a hint of legs and tail.

Roger Knutson found problems with getting photographs for the book. There was a real risk that the photographer would

DIE HARDER!
The Armadillo way

ARMADILLO PENIS RESEARCH

In northern Florida, the roadside is littered with the bodies of dead nine-banded armadillos. Dr Diane Kelly of Duke University in Durham, North Carolina made an amazing discovery.

She looked closely at the penises of these dead armadillos. An erection happens in a penis when the blood flows in, but can't easily get out. The fluid pressure inside the penis increases. This pressure pushes against a rigid sheath of fibres that run along the length of the penis. But exactly how are the fibres arranged?

Previously, scientists thought that the fibres were arranged in a spiral that wound around on the inside of the penis, a bit like a spiral staircase. But Dr Kelly discovered that (at least in the nine-banded armadillo) there are two sets of spirals at right angles to each other — a much stronger arrangement than just a single spiral.

Dr Kelly thinks that it is possible that other mammals may have this twin-spiral arrangement, but that nobody has noticed.

end up as flattened fauna. So they had one person take the photographs, and two other people look out for the cars. He warns of the dangers of trying to make an exact identification, 'Better to know that a particular specimen is *possibly* a vole, than to be certain it is *Microtus pennslvanicus*, and join it on the road.'

Roger Knutson also offered some handy hints. He says that if you wish to become a collector of flattened fauna, it's safer to collect only pictures, not samples. You're better off leaving the 'fur pies' on the road, because they may have fleas or rabies. He also advised that accurate identification is sometimes not possible. For example, a crushed armadillo often looks just like a squashed cheap muffler.

Some Humans are Mean

Another researcher in this field is Professor David Shepherd from the Southeastern Louisiana University. He and his colleagues did an experiment. They placed eight fake turtles and snakes on the road, so that they were directly in the path of the wheels of the vehicles. They also put other fake turtles and snakes way off to the side of the road, where the drivers would actually have to make an effort to run over them.

They found that we humans are actually a fairly nice bunch — 87 per cent of drivers tried to avoid the turtles and snakes. Another 7 per cent of drivers didn't notice the fakes, and accidentally ran over them.

But 6 per cent of drivers had a mean streak. Shepherd said 'A truck driver even crossed the center lane, went into the opposite lane of traffic and drove onto the shoulder of the road to run over a "turtle". And a policeman crushed a snake with his

tyres and then stopped and pulled out his gun. I quickly jumped out from some bushes and explained it was a fake.'

Rapid Compression and Solar Dehydration

Another researcher in the field of flattened fauna was William E. Artz, from the Department of Food Science at the University of Illinois. He wrote for the *Annals of Improbable Research*, a most excellent comedy magazine for scientists. His paper is called 'Drying by Rapid Compression and Solar Dehydration'.

In his paper, he talks about how the tyres of big 18-wheeler trucks squash animals on the road, allowing the sun to dry them out nicely. He writes in pseudo-scientific language. He says the equipment used was a 'rapid compression system followed by solar drying on a large flat surface — usually referred to as a "road kiln". The best designs utilise multiple effect systems. For example, a compression system with nine circular compression devices (approximate diameter = 1 metre) on each side of the vehicle is often used. A singular circular compression device, followed by three to four dual sets of compression devices, is the typical design. The device is applied to the sample (optimum sample mass range = 1–3 kilograms) over a very short time period (0.1–0.2 seconds). A GVW (Gross Vehicle Weight) of > 15 000 kg

DUTCH HEDGEHOG ROADKILL RESEARCH

The 'standard' truth about why hedgehogs died on the road was simple — the warm road attracts insects, the hedgehog comes to eat the insects, and when a car approaches, they foolishly roll up into a ball rather than try to run away.

But a study by the Netherlands Society for Mammology and the Protection of Mammals disagrees. J. L. Mulder wrote the study, called *Hedgehogs and Cars, An Analysis of the Literature.*

First, the hedgehogs don't look for food on the road — they are simply trying to cross the road. Unfortunately, they usually move around dusk — which happens to be rush hour for human society.

Secondly, they will roll up into a ball only when they are directly attacked — not when a car approaches. They will sometimes 'freeze' in paralysis, but this is not common. Most frequently, they will try to run. The reason that they get run over is that they simply are not fast enough.

enhances first stage process efficacy and is strongly recommended.'

In his paper, he has one graph that shows the moisture changes that happen as a result of the First Stage Rapid Compression (when the animal is first hit, and loses blood and other precious bodily fluids) over a few seconds. The second graph shows the moisture changes that then happen as it dries on the road (over the next few weeks) thanks to the heating effects of the sun. The third diagram shows 'a finished, compressed, dried sample' — a dead skunk.

Project Roadkill

But in the rich field of research known as flattened fauna, Dr Splatt has a unique approach.

His real name is Brewster Bartlett, and he teaches science at the Pinkerton Academy in New Hampshire. The National Science Foundation sponsors his Roadkill Project (http://earth.simmons.edu/roadkill/roadkill .html). Some schools have participated for four years. In this project, 3000 or so students from 15 states across the USA go onto the roads looking for flattened fauna. Each school counts how many of each different type of animal were killed in its 'section' of road. Altogether, they find about 3000 to 6000 dead animals each year, ranging from squirrels to birds and deer and moose. So far, they have identified about 30 different animals that have been flattened by road traffic. They will then feed their data to the central collection area, and sometimes show their data on the web.

But sometimes the animal is so flattened, that it can't be recognised. In these cases, they call it a URP — an Unidentified Road Pizza.

One typical data collection (http://rjh. vitts.com/EnviroNet/Roadkill.htm) over

ROADKILL CUISINE

On the internet, the Roadkill Café site (http://www.pressanykey.com/ roadkill.html) opens its menu with the words, 'You kill it, We grill it' and 'eating is more fun, when you know it was hit on the run'.

The entrees include Center Line Bovine, The Chicken that Didn't Cross the Road, and Flat Cat.

Under Bag'n'Gag (the daily take-out lunch special), it offers 'Anything dead . . . in bread'.

In the main course (A Taste of the Wild Side, Still in the Hide) it offers Swirl of Squirrel, Smidgen of Pidgeon, Chunk of Skunk and Rigor Mortis Tortoise. The speciality section (Canine Cuisine — You'll Eat Like a Hog, When You Taste Our Dog!) lists Poodles N Noodles, Collie Hit by a Trolley, German Shepherd Pie and Round of Hound. An Australian version might include: Emu Stew, Koala Kola, Kangaroo Carapés, Enchilada Echidna, Snake Steak or Wombat Wellington!

eight weeks counted four birds, five chipmunks, nine frogs, four mice, one muskrat, one porcupine, two rabbits, six raccoons, two rats, two skunks, eight Gray Squirrels and two URPs. The students relate the killing rates to (among other factors) the phase of the moon, and the temperature.

The Gray Squirrel has consistently been the most killed animal on the Project Roadkill database. Now that Project Roadkill have discovered this, they are trying to find exactly what it is that makes this animal such a popular roadkill — and maybe work out a strategy to reduce the deaths.

The Roadkill Cookbook

David Larcey has taken a rather special approach with his book called *The Roadkill Cookbook*. This is a collection of exquisite recipes, that have for their main ingredient some kind of flattened fauna that has been peeled, or scraped, off the road.

Luckily, the state of Virginia has stepped in to help these do-it-yourselfers, by legalising 'take-home roadkill'. Previously, only the state had the right to remove roadkills from the road. But it cost the state money to remove the carcasses from the road. It wanted to save money, so it changed the law. It is now legal (but not necessarily safe or tasteful) in Virginia to take a roadkill home and eat it, after you have run it over.

KILLER WINES FOR ROADKILL CUISINE

John McEndry wrote a very funny article with the above title for *Smart Wine Magazine* (http://smartwine.com/consumer/swaug96/sw089607.htm). He gives advice about selecting the right wine for the right roadkill.

He first showed his expertise by realising that different environments gave added flavours and textures to the roadkill. So while the roadkill on a busy multi-lane highway would reek of rubber, oil and brake fluid (and be very flat), the roadkill from a quiet country dirt road would have fewer hydrocarbons, more 'texture' (ie dirt), and have more of the third dimension (because of fewer vehicles squashing it).

So the very first step was to realise this difference, and to prepare the roadkill appropriately. For example, a roadkill rich in hydrocarbons would be very easy to sauté, while one that was rich in gravel would need extra cleaning.

If you could appreciate the roadkill for what it was, the choice of wine was almost intuitive. So while a squashed chicken (Peterbilt Poulet) would obviously go with a Chablis Ordinaire, Secondary Road Possum would need a strong red wine to compensate for any added flavours (such as road tar, anti-freeze or creosote) that might be carried in the meat.

But McEndry's genius flowered when he discussed what to do with leftovers. After all, you couldn't simply throw them back out in the street. He recommended recycling the leftovers by cooking them in a big stew (called Head-On Stew) for eight hours, and serving them in big hub caps. The perfect accompanying wine? Just go to your local vineyard tasting room, and get them to save you a few buckets of Dump Bucket Burgundy!

Enroadment is Similar to Fossilisation

However, a geologist has taken a completely different approach to flattened fauna. Hans Machel from the University of Alberta realised that there is a lot of similarity between the 'process of fossilisation', and the 'process of decay' that happens in an animal that has undergone the 'enroadment' experience. He's been taking photographs of roadkills at various time intervals — minutes to weeks after they were killed.

There are significant differences. In fossilisation, the animals get taken to pieces by other living creatures, or by the action of the wind and the waves, or by natural chemical and biological decay. In fossilisation, the flattening happens by the dead animal being buried under ever-increasing weights of sediment — not by repeated tyre impacts.

Even so, Machel has found a few useful educational similarities between fossilisation and the decay of a roadkill.

RABBIT PLAGUE

On one of our Outback trips, we ran into a rabbit plague that stretched 10 kilometres. We were slowly trundling west toward Broken Hill, in the half-light immediately after sunset.

There were rabbits all over the road, and onwards to the horizon as far as the eye could see.

When we drove at a walking pace so that we could miss them with the front wheels, they dived under the back wheels — and then they made a terrible splatting noise as they went to Rabbit Heaven, courtesy of the 8.25 x 16 Supergrippers (each carrying over one tonne of Volvo C304). If we kept on going at a cruising speed, we would run over rabbits with both the front and back wheels. There was just no way to avoid them unless we came to a complete stop. And no we didn't have rabbit for dinner around the camp fire!

The Field Expands to Birds and Insects

Peter Hansard and Burton Silver have written a book called *Flattened Fauna* which, like Roger Knutson's original work, will help you identify an animal after it has been compressed and dried.

This book was so successful that they've written another book called *What Bird Did That?*. You can quite accurately identify the birds by looking at what they have dumped on your shiny paint (but only if they're American birds). Unfortunately, according to Amazon.com (an on-line bookstore), the European equivalent, *The Comprehensive Field Guide to the Ornithological Dejecta of Great Britain and Europe*, is out of print. If you want a copy, you will have to look through second-hand bookshops.

Their next book in the series will be *What Insect Was That?*. Once again, by looking at what's left of the insect on your windscreen, you can identify what insect it used to be.

That Gunk on Your Car

But they've already been beaten in the field of flattened insects. In 1996, Mark Hostetler wrote a useful book called *That Gunk on Your Car*. If the sloppy yellow streak on your windscreen is about 2–3 cm long, then it's probably a butterfly of the tiger swallowtail variety, *Pterourus* (or *Heraclides*) *glaucus*. But if you have a little greyish dot with a minuscule drop of red blood in the middle, you've probably got *Aedes sollicitans*, also known as the female salt-marsh mosquito.

Mark Hostetler wrote this while he was doing his Ph.D. in zoology at the University of Florida, in Gainesville. He collected his specimens from his own car (a Honda Accord), and from the buses at the local Gainesville Greyhound Depot. He has some two dozen illustrations in his glovebox-sized book, combined with the standard field guide description of these insects — what the insects look like, how they are born, live and die, and their behaviour.

He did notice that different insects had rather different behaviours — which would lead to different patterns of death.

Take lights, for example. We all know that moths love headlights, because they confuse them with the moon, which they use for navigation. But monarch butterflies adore those yellow safety reflectors in the middle of the road.

Lovebugs usually deposit their eggs in dead and decaying animals or plants. They get attracted by the fumes. For some unknown reason, lovebugs also love some of the fumes that come out of car exhausts, after the fumes have been changed by sunlight. They get tricked, and will often dump their eggs on busy streets on sunny days. While they're on the road, they are often killed by traffic.

Mark Hostetler was so thorough in his work, that in 1997, the *Annals of Improbable Research* awarded him one of its Ig Nobel Prizes. These prizes honour scientific, medical and linguistic achievements that 'cannot or should not be reproduced'. He won the Entomology Prize, and appeared on the *Tonight Show*, where he helped Jay Leno and Drew Carey identify random insects' splatts.

Angry Eagles

When there are flattened fauna on the roads, then other creatures will come along to have a feed at the Roadkill Café. It's been this way since around 1910, since cars became a significant force on our planet. But recently, we humans have been interfering with the delicate balance that we set up back in 1910.

Wedge-tailed eagles have enjoyed eating live rabbits ever since rabbits were introduced into Australia in 1859, by Thomas Austin. But the recently introduced Calicivirus has been killing rabbits all across Australia. Because of the shortage of rabbits, the big eagles had to join the other animals at the Roadkill Café.

The problem is that wedge-tail eagles are pretty large — wing span up to 2.5 metres (8 ft), weight up to 5 kg (11 lb) — so they're a bit slow to get out of the way of a speeding vehicle. Recently on the Barrier Highway near Broken Hill, a baker accidentally hit an eagle with his van. The eagle came straight through the windscreen and began to tear the cabin apart as it desperately tried to get out. The driver immediately had to abandon the vehicle to avoid being hurt.

Kangaroos

In the Outback, there are many roads with gentle undulations. Condensation and moisture settles in these dips after sunset. Kangaroos come to these dips to gather moisture at the side of the road. When motorists come speeding along this slightly undulating road, at night, at high speed, and with the high beam on, they sometimes come over a small ridge to find a kangaroo in their path. If the kangaroo is unlucky enough to look straight into the headlights, it will be blinded, hypnotised and, finally, killed.

In all my driving, I have run into only one biggish animal. I was driving into the sunset, when a kangaroo suddenly jumped from behind a bush, in front of my 4WD. I saw it just before it hopped in front of the passenger-side bullbar. Even though the kangaroo was much lighter than my four-tonne Volvo C304, and I was travelling at only 30 kph, all the lights immediately stopped working. I tried, but I simply could not get the lights working again.

I stopped there at the side of the road for the night, and in the morning, the lights worked just fine. No matter how hard I looked, I never could find anything wrong with them.

REFERENCES

William E. Artz, 'Drying by Rapid Compression and Solar Dehydration', *Annals of Improbable Research*, Vol. 1, No. 4 July/August 1995, p 28.

D.A. Kelly, 'Axial Orthogonal Fiber Reinforcement in the Penis of the Nine-banded Armadillo (*Dasypus novemcinctus*)', *Journal of Morphology*, Vol. 233, No. 3, September 1997, pp 249–255.

Laurence A. Marschall, (book review) 'Flattened Fauna: A Field Guide to Common Animals of Roads, Streets, and Highways', *The Sciences*, May/June 1988, p 51.

—— (book review) 'That Gunk on Your Car', *The Sciences*, January/February 1997, p 47.

New Scientist, No. 2065 (Feedback), 18 January 1997, p 56.

'Why Armadillos Die Harder', *Fortean Times*, March 1988, p 6.

http://earth.simmons.edu/roadkill/roadkill.html

http://rjh.vitts.com/EnviroNet/Roadkill.html

http://www.pressanykey.com/roadkill.html

http://smartwine.com/consumer/swaug96/sw089607.html

MELATONIN – MAGIC OR MADNESS?

If you hang around health-food stores, you must have heard of melatonin, currently the most fashionable hormone in the entire known universe. In California in 1995, melatonin sales beat aspirin sales! Melatonin is 'The Next Big Thing' in popular medicine.

In the USA, melatonin is classified as a food additive (because it's found naturally in tomatoes and bananas), so there are no legal restrictions on its sale. It's claimed that melatonin will cure everything from cancer to jet lag; make your periods more regular; strengthen your immune system; rejuvenate your sex life; fix up your heart; make your muscles stronger; help you get a good night's sleep and even slow down the ageing process!

THE THIRD EYE

We are still learning about the pineal gland, but so far we have learnt that it has different functions in different creatures. It has long been called 'the third eye'.

In the 'lower' (such as fish) vertebrates (creatures with a spine), the pineal gland is sensitive to light, and even has some of the structures you find in an eye. It is a 'primitive' eye.

In the 'higher' vertebrates (amphibians and higher), the pineal gland has two functions: an organ to sense light (if not images), and a source of hormones. In some reptiles, the pineal gland is directly exposed to sunlight through a translucent membrane on top of the reptile's head. In some vertebrates, the pineal gland is quite a sophisticated 'eye', and can probably 'see' images, not just light.

In mammals, it has no function as an 'eye', because it is buried deep in the brain, away from external light. The pineal gland receives electrical signals, via the retina, about light. It releases chemicals such as somatostatin, noradrenaline (norephinephrine), serotonin (famous for its levels being increased by Prozac), histamine and melatonin.

Unfortunately, there's not a lot of evidence for any of these claims.

But that hasn't stopped people from writing sensationalist and noncritical books such as *The Melatonin Miracle* by Walter Pierpaoli and William Regelson, *Stay Young the Melatonin Way* by Steven Bock and *Melatonin: Nature's Sleeping Pill* by Ray Sahelian. However, Josephine Arendt has written an extremely clear, well-balanced and sensible book called *Melatonin and the Mammalian Pineal Gland*.

Melatonin and the Pineal Gland

Melatonin gets its name from the Greek word *melas*, which means 'black'. It is also called *5-methoxytryptamine*. The pineal gland makes melatonin by turning an amino acid, *tryptophan*, first into serotonin, and then into melatonin. Melatonin acts on the skin cells called melanocytes, which in turn produce a chemical called melanin. This is the chemical which gives tanned skin its dark colour.)

Melatonin is a hormone made in the pineal gland. It is tiny and white and is situated deep in the brain. It's about 0.65 cm (quarter of an inch) long, and weighs about 0.1 gram (0.0035 oz). It's called 'pineal' after the Latin word 'pinea' which means 'pine cone', because the gland actually looks like a tiny pine cone.

History of the Pineal Gland

The Greek anatomist Herophilus (335–280 BC) knew of the pineal gland. He believed that it was some kind of 'valve' that controlled the flow of 'thought' which was stored in bins higher up, in the lateral ventricles of the brain.

In 1644, in his *Principles of Philosophy*, the philosopher René Descartes (1596–1650) wrote that the pineal gland was the 'seat of consciousness' in the brain. He thought that because the pineal gland was the only single (not double, or paired) organ in the brain, it *must* be the link between the mind and the body. Even today, we still don't know where 'consciousness' resides — we *think* that it is somewhere in the brain, but we really don't know. Descartes is famous for the saying, '*I think, therefore I am*'. (The scientists' version of this is, '*I think, therefore I get paid*'.)

Descartes thought that the eyes sent information about the real world via 'strings' to the pineal gland. The pineal gland would then send 'humours' down to muscles through hollow tubes, which would make us respond to what we saw. It took medical science three centuries to show that he was surprisingly close to today's view about the pineal gland. Today, we see his 'strings' as nerves, and 'humours' as hormones.

Melatonin - Hormone of Darkness

Melatonin is not released continuously, 24 hours per day. Scientists sometimes call it 'The Hormone of Darkness', because it is released soon after your environment becomes dark.

This release of melatonin stops as soon as the light level gets above 2500 lux. One lux is the amount of light cast by a single standard candle onto a surface from a distance of one metre. The average night-time house lighting is 250–500 lux, a dull day in northern Europe is about 10 000 lux, while a bright day on the equator is about 80 000 lux.

THE DANGER OF L-TRYPTOPHAN

In the 1970s, it was discovered that microscopic doses of L-tryptophan might help people sleep. It became popular in health-food stores.

In 1989, a Japanese company accidentally made and sold contaminated L-tryptophan. This impure preparation created a new syndrome (the eosinophilia-myalgia syndrome) which killed 45 people, and left thousands of others with a lifetime of pain. Even today, we're not sure exactly what the dangerous contaminant was. In 1990, L-tryptophan products were taken off the American market.

Perhaps the same sort of thing could happen with melatonin. After all, it is not made to the same quality control standards that pharmaceutical companies have to pass. In 1995, in the USA, the Committee for Product and Label Integrity found that 25 per cent of melatonin products did not pass their test.

Blood Levels of Melatonin

Melatonin in your bloodstream is rapidly destroyed by your liver, with half of it vanishing in about 45 minutes. 'Typical' blood levels of melatonin are 10 picograms/ml (pg/ml) in the daytime, and 100 pg/ml at bedtime (a picogram is very small — one millionth of one millionth of a gram).

However, these are the *natural* levels, in response to the melatonin that is made in the pineal gland, and dumped directly into the bloodstream.

The situation is quite different when you put a melatonin tablet in your mouth, and it travels to your stomach, and then gets absorbed into the bloodstream. One study looked at healthy young men swallowing 0.5 mg melatonin tablets. Their peak melatonin plasma levels varied enormously by a factor of 20 — from 480 to 9200 nanograms/litre. Two people taking the same dose of melatonin can have enormously different blood levels. You have to wonder just how much the melatonin levels vary in the millions of people who take it every day.

You make heaps of melatonin in childhood and adolescence, less in adulthood, and much less again in old age. Typical peak night-time melatonin levels are 125 pg/ml at age 6, 75 pg/ml at age 20, 50 pg/ml at age 50, and 30 pg/ml by the age of 80. In fact, the pineal gland shrinks and becomes calcified as you get older, and this calcium can be easily seen in X-rays of the skull. Some people claim that if you take melatonin, you can become young again.

Discovery of Melatonin

Melatonin was officially discovered in 1958, by Aaron B. Lerner and his colleagues at Yale University. Over a four-year period, he collected some 200 000 pineal glands of cattle from the Armour & Co. meat processing company. The dried pineal glands were so tiny that all 200 000 of them would fit into a couple of shoeboxes. From this, he extracted microscopic samples of 100 micrograms (0.0000035 oz) of melatonin. This amount was too small to see, but he and his team were able to analyse it, and then make it artificially.

Most of what we know about melatonin applies to animals such as mice. We don't yet know how melatonin 'locks', or entrains, the cycles of the various hormone systems in humans, and we have a very poor understanding of exactly where melatonin acts in the human brain.

Uncontrolled Melatonin Trials

At the moment, we have the bizarre situation where melatonin is the first drug ever to have its clinical trials on the internet — and these trials are uncontrolled. They are not supervised by scientists, and the data is not collected by statisticians. People have formed groups where they take melatonin and then discuss what happens. In about 10–15 years, we'll know what some of the side effects of melatonin are, thanks to the millions of unsupervised people taking it.

If the people taking melatonin were monitored more closely, we would probably discover the possible long-term effects sooner, and maybe fewer people would suffer.

Dr Thomas Wehr is a psychiatrist at the National Institute of Health. He has been doing research into biological clocks for over a decade. He was quoted in *Discover* magazine as saying, '*The public has gotten way ahead of the researchers. The experiments are being done on a massive scale by people just taking melatonin without controls.*'

Melatonin Sleep Cycles

First, we have found that melatonin is (to some degree) involved in regulating your sleep cycles. Melatonin lowers your body temperature, and so makes you a little bit sleepy. However, according to Professor Drew Dawson at the Centre for Sleep Research at Adelaide University, ironically melatonin has its strongest effect on those people who don't really need it — young adults who have no sleeping problems.

In the mid-1980s, scientists discovered that humans and rats that were kept in total darkness gradually drifted out of their normal 24-hour day into a 25-hour day. But when the rats were given melatonin, they went back to a 24-hour day,

even in total darkness. So we think that, at least in rats, melatonin could re-adjust the body clock.

Melatonin can re-set the body clock of people who are blind. Some people who are totally blind still make melatonin. They tend to sleep well. But some blind people make no melatonin at all. They tend to have a disturbed and broken sleep. Melatonin can help these people have a better sleep.

We still have much to learn about the effects of melatonin on people.

Consider the experiment in which 16 men (in two separate groups) were kept in dimly lit rooms, and were given melatonin at 8 pm. The experiment was run by chronobiologist Benita Middleton and her colleagues, at the University of Surrey in Great Britain. They found that different men had different biological clocks. Some of them reached their peak temperature later in the day. These men responded badly to the melatonin, and had very fragmented and restless sleep.

Melatonin
Recommended
by Rats
500 mg

NEUROLOGY OF THE PINEAL GLAND

It was only as recently as 1965 that the nerve supply of the pineal gland was well described.

The two retinas in the eyes send the processed electrical signal down the optic nerve to the hypothalamus, deep in the brain. In the hypothalamus, the SCN (SupraChiasmatic Nucleus) monitors the length of the day, and controls the release of many chemicals — including melatonin. (Some scientists think that the SCN is the 'master clock' that runs the human biological clock.) Nerves run from the hypothalamus to the pineal gland.

Some of these nerves are part of the sympathetic nervous system. The sympathetic nervous system can set off a 'flight, fight, fright or making love' response.

Other experiments show that if you choose to sleep when your natural melatonin level is at a peak, you will have a restful sleep. It seems that the timing of when you take melatonin is critical. How many of the millions of people who take melatonin every day stop and think about *when* they are taking it?

Melatonin and Sexual Maturity

Secondly, we know that melatonin also controls (to some degree) the changes that happen during the process of sexual maturity in adolescence. The presence of melatonin helps slow the onset of sexual maturity. Some sex hormones (such as luteotropin) appear only after the levels of melatonin have dropped.

Children with tumours of the pineal gland (and who make very little melatonin) reach sexual maturity very early. On the other hand, young men with very high levels of melatonin have delayed puberty. High levels of melatonin force the anterior pituitary gland to release high levels of prolactin, a substance which depresses the male sex drive.

What will be the long-term effects on the sex drive of the millions of men who take melatonin?

Melatonin and Jet Lag

Thirdly, melatonin is also involved in your regular daily rhythms.

Many people have suffered jet lag. If you allow the human body clock to run at its own rate, it will shift to a 25-hour day. When you fly to the west, your 'day' will be longer — flying from New York to San Francisco, or from Sydney to Perth, will lengthen your 'day' by three hours.

In general, you will adjust more easily to jet lag when you fly to the west. You usually adjust naturally by about one hour per day. So it will take your body three days to adjust to a three-hour time shift .

In the case of jet lag, melatonin supposedly works by 're-setting' your body clock, which is thought to be in your hypothalamus. The research into jet lag shows mixed results. It may turn out that melatonin will help only a few types of jet lag (for example, when the trip is to the east, and the 'time slip' is between two and four hours).

But recent research has found that hamsters (we don't know about humans, yet) have another body clock — in their retina. This retinal body clock is also sensitive to melatonin. So what does added melatonin do to this other body clock, if it turns out to exist in humans as well? We simply have no idea at the moment.

One thing we do know is that the melatonin story will turn out to be far more complicated than what the populist books are telling us.

Sleeping Pill

Fourthly, melatonin will supposedly help insomniacs sleep better. Some studies (usually on a very small scale) claim that elderly people with sleeping problems sleep much better if they take a little melatonin.

Dr James Jan from the Children's Hospital, in Vancouver, British Columbia found that melatonin in doses of 2.5 to 10 mg helped children with definite medical conditions (such as epilepsy, autism, Down's Syndrome or Trisomy 21, and cerebral palsy) get to sleep.

But these children had definite medical problems. Should people with no real problems take a hormone such as melatonin for many years?

Anti-Cancer

It is still early days and the results haven't yet been proven, but it seems that melatonin might be able to help in the treatment of breast cancer, when it's given with the standard anti-cancer drugs. However, the anti-cancer effect (if it's real) is not very strong.

Melatonin also acts as an anti-oxidant and a scavenger of free radicals — but only at very

EYE AND SLEEP EXTREMES

According to the *Guinness Book of Records*, the eye muscles move more than 100 000 times per 24-hour day. A lot of these eye movements happen during dreaming — in REM (Rapid Eye Movement) sleep.

There is an extremely rare disease called Chronic Colestites (total insomnia), in which the sufferers can go many years without 'regular' sleep. Around midnight, they may put on night clothes, go to bed and read a book for an hour, get up again and get dressed, and then start a new 'day' of activity.

In some parts of the world, junior hospital doctors *have* to work very long hours. In June in 1980, Dr Paul Ashton, 32, (anaesthetics registrar at Birkenhead General Hospital, Merseyside, United Kingdom) worked for 142 hours in one week. He said that the week was *'particularly bad but not untypical'*. Sleepy doctors must be more likely to make mistakes than doctors who are well-rested.

MELATONIN AND POISONS

A few experiments seem to show that melatonin can fight nasty chemicals.

In one experiment, white blood cells were exposed to melatonin and then blasted with ionising radiation. They suffered 70 per cent less damage than white blood cells that had not been previously exposed to melatonin.

In another experiment, rats were injected with melatonin, and then given the nasty chemical, paraquat, which usually causes severe lung damage. The melatonin seemed to give them some protection.

Perhaps in the future, melatonin might be useful in fighting a short-term insult such as radiation or poison.

high blood concentrations (much higher than what's normally found in the blood).

Some studies show that in animals melatonin can stimulate parts of the immune system. But there have never been any controlled studies on humans. And one real worry is that melatonin could trigger an underlying auto-immune disease that you didn't know you were likely to get, such as rheumatoid arthritis.

Melatonin and the Heart

Fred W. Turek from Northwestern University is critical of some of the claims for melatonin. He says that 'not one study' supports the claim by Regelson and Pierpaoli (in their book, *The Melatonin Miracle*) that melatonin can in any way prevent heart attacks.

Contra-Indications of Melatonin

There are some cases in which melatonin is specifically forbidden. Victor Herbert and Ruth Cava from the American Council on Science and Health warn that pregnant women, breastfeeding mothers, children, and people with immune-system disorders should not take melatonin — simply because we don't know what it does to these people. People with hormonal disorder diseases (such as diabetes) shouldn't take it either.

Hype = Bad Science

The books that hype melatonin use three main tactics in order to give the general public an unbalanced and non-scientific understanding of melatonin.

First, they exaggerate the importance of a few minor studies, that are usually done with animals, not humans.

Secondly, they claim that certain theories are facts, when they are really just theories, with little or no supporting data.

For example, *The Melatonin Miracle* claims that melatonin will lower blood pressure, lower the levels of blood cholesterol and protect you from heart attack and stroke. Yet the authors do not quote one single study in humans that backs up their claim. However, the book does refer to a friend of one of the authors who took 5 mg of melatonin. After a few months, this friend claimed to have normal blood cholesterol levels. They don't tell us anything else about this friend — did he change his diet too or did he take anything else to lower his cholesterol? This is an *anecdote* (not a clinical trial) about *one* person (not a statistically useful number such as 100) and is not scientific proof!

Another claim in *The Melatonin Miracle* is that the pineal gland *controls* daily rhythms and sexual maturity in humans. The fields of study of daily rhythms and sexual maturity are huge and complicated, and practically every scientist in these fields disagrees with such simplistic statements.

And thirdly, Turek suggests that they always conveniently ignore data that disagrees with their pet theories.

Fundamentally Flawed Experiment

Take, for example, the famous experiment organised by Dr Walter Pierpaoli, one of the authors of *The Melatonin Miracle*. According to the melatonin supporters, this experiment supposedly proved that melatonin could make humans young again.

Now remember that the pineal gland makes melatonin, and that as you get older, you make less melatonin.

Pierpaoli got ten young (four-month-old) laboratory mice, and ten old (18-month-old) laboratory mice. Then he got a neurosurgeon to transplant the pineal glands from the old mice to the young mice, and vice versa.

And lo and behold, the young mice started ageing rapidly, while the old mice regained some of their youthful vigour. In fact, the young mice with the old pineal

glands lived one-third *less* than average and the old mice with the young pineal glands lived one-third *longer* than average. It sounds like perfect proof that melatonin restores your youthfulness, and that lack of melatonin makes you older.

Pierpaoli claimed, as a result of this experiment, that taking small amounts of melatonin will let a human live to 120 years.

There are a few problems with this experiment.

First, mice are not humans.

Secondly, 20 is too small a sample size on which to base any real statistics.

The third problem is very serious, and makes Pierpaoli's claim totally ludicrous. Steven M. Reppert and David R. Weaver from the Harvard Medical School pointed out that Pierpaoli didn't actually measure the melatonin levels in his 20 mice. He just *assumed* that the melatonin levels changed. But it turns out that the particular strains of mice he studied don't make melatonin at all, ever — thanks to a genetic defect.

There is a lot of difference between mice with a genetic defect that stops them from making melatonin, and humans who naturally make melatonin.

Astonishingly, the book *The Melatonin Miracle* uses this pathetic 'study' to claim that humans can live 30 per cent longer if they take melatonin.

Melatonin Controlled in Australia

In the USA, melatonin is uncontrolled. People can take as much as they want. But in Australia, melatonin was classified as an S4 drug on 19th December 1997. This means that if Australians want to get it legally from overseas, they need a prescription from their local doctor.

Melatonin cannot be patented, so there is no financial incentive for a pharmaceutical company to run the enormously expensive clinical trials to see exactly what the effects of melatonin on humans are. It would simply cost them tens of millions of dollars.

If you think that melatonin can help you, you can increase your natural melatonin levels without taking the tablets. Cut down on the number of calories that you eat, and have only two or three meals per day, rather than grazing on lots of snacks. You can get a further increase by having regular exercise, and by dimming the lights at night. Also, avoid alcohol and caffeine.

So if you want to be safe you should avoid taking melatonin, at least until we know what the effects are.

And if you like to believe in conspiracy theories, maybe the giant human melatonin experiment is being secretly carried out by super-smart mice.

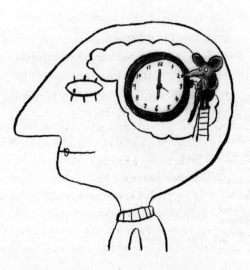

TOO MUCH MELATONIN

Richard J. Wurtman is the Cecil H. Green Distinguished Professor in Brain & Cognitive Sciences at the Massachusetts Institute of Technology. He has been working with melatonin for a long time. In 1963, he wrote one of the early papers discussing melatonin's role as a hormone.

In the early 1980s, in short-term clinical controlled trials, he was giving volunteers massive doses of melatonin of up to 240 mg. The patients didn't sleep at all well taking these quantities. Today he conducts clinical trials with doses as low as 0.1 mg. Taking amounts in the 0.1–0.3 mg range, results in similar melatonin blood levels to those that occur naturally. Professor Wurtman is horrified by the millions of people who daily take 10 mg, and more, of melatonin. Some of these people consistently have melatonin levels in their blood 30 times greater than normal.

He was quoted in *Science News* as saying, '*At present, all of the health-food store "melatonin" preparations I've seen contain amounts of the drug that are far too high and that raise melatonin concentrations in the blood far above any that occur normally*'.

This is a worry. We already know that some glands stop producing their natural hormone if the levels of that hormone get too high. Will taking melatonin tablets force the pineal gland to stop making melatonin?

Melatonin is extraordinarily safe in terms of having an overdose, at least in the short term. Volunteers have taken 6000 mg of melatonin every night for a *month*! They suffered short-term side effects of stomach discomfort and tiredness, but no long-term side effects were reported.

But what happens to somebody who takes melatonin for *ten years*?

REFERENCES

Geoffrey Cowley, 'Melatonin', *Newsweek*, 14 August 1995, p 46–49.

Charles Czeisler et al, 'Suppression of Melatonin Secretion in Some Blind Patients by Exposure to Bright Light', *New England Journal of Medicine*, Vol. 332, No. 1, 5 January 1995.

Atholl Johnston et al, 'Variable Bioavailability of Oral Melatonin', *New England Journal of Medicine*, Vol. 336, No. 14, 3 April 1997.

Aaron B. Lerner, 'Hormones and Skin Color', *Scientific American*, July 1961, pp 98–108.

Dr Sarah Mahoney, 'Waking Up to Melatonin Use — Why doctors need to be prepared', *Australian Doctor Weekly*, 6 February 1998, pp 35–38.

W. Pierpaoli, Lesnikov V.A., 'The pineal aging clock. Evidence, models, mechanisms, interventions,' *Annals of the New York Academy of Sciences*, Vol. 719, 31 May 1994, 461–73.

S. M. Reppert, David R. Weaver, 'Melatonin Madness', *Cell*, 29 December 1995, pp 1059–1062.

Fred W. Turek, 'Melatonin Hype Hard to Swallow', *Nature*, Vol. 379, 25 January 1996, pp 295–296.

Richard J. Wurtman and Julius Axelrod, 'The Pineal Gland', *Scientific American*, July 1965, pp 50–60.

NAME YOUR OWN ELEMENT

Immortality Through Names

We all want our little bit of immortality, but a lot of us are happy to achieve it by having children. Some of us get our taste of immortality by putting our names on buildings, or even on larger things like mountains. The trouble is that there are squillions of buildings and mountains, so it's a fairly ordinary kind of immortality.

If you want to be really remembered, get your name on an element. Buildings and mountains will eventually crumble to dust, but an element is forever. By mid-1998, we had discovered only 112 different elements in the entire known universe. All the matter in the known universe is made up of various combinations of these 112 elements. Glenn Seaborg eventually got his name on a radioactive element (Seaborgium), but he had to overcome a major objection by some of his fellow chemists — they didn't like the fact that he was still alive!

Elements

What exactly is an element? Chemists say that it's any substance that cannot be broken down into simpler substances by ordinary chemical and physical reactions.

If you heat iron ore (which is a combination of the elements iron and oxygen) under the right conditions, you can isolate out the iron. But no matter what you do to the element called iron (heat it, freeze it, pound it with a hammer, add other chemicals and so on), you cannot break it down into simpler substances.

If you look around, you can see a few elements. Copper (in plumbing pipes) is an element. Lead (used for flashing on roofs) is also an element. Iron (in carpenters' nails) is another element. Gold, silver, and carbon (which has a special form called diamond) are all elements too. The mercury in a thermometer is another element.

But steel is not an element, because steel is made of two other elements, iron and carbon. Glass is not an element, because it's made from two other elements, silicon and oxygen. Most of our atmosphere is made from just two elements, nitrogen (about 80 per cent) and oxygen (about 20 per cent). Most of our food is made from only four elements — hydrogen, oxygen, carbon and nitrogen.

ATOMIC NUMBERS

An atom looks a little bit like a mini-solar system.

There is a central core (a bit like the Sun), which is called a nucleus and there are a whole lot of electrons (a bit like the planets) whirling around the nucleus.

The electrons have a negative charge. In the nucleus are one or more protons, which have a positive charge. There are also some neutrons in the nucleus. They have no charge. There are roughly as many neutrons as protons in the nucleus.

The 'atomic number' of an atom is the number of protons. So hydrogen, which has only one proton, has an atomic number of one. Helium, which has two protons, has an atomic number of two. Uranium, with 92 protons, has an atomic number of 92.

Of course, nothing is simple.

This neat little 'solar system' picture is not really that accurate, especially with the heavier elements. With lead–198, two of the protons and four of the neutrons are not in the nucleus. They are actually outside the nucleus, and orbit or 'skim' the nucleus a bit like low-flying satellites orbit our Earth!

Classifying Elements

There are many ways to classify the elements.

For example, we can classify elements by their 'state of matter'. So, at the temperature and pressure we enjoy in our houses, two of the elements are liquids, 11 are gases, while the rest are solids.

A very common way to classify elements is to divide them into metals and non-metals. And another way to classify the elements is into non-radioactive and radioactive elements.

Radioactive Elements

'Non-radioactive' just means that the element is stable, and will last 'forever' (or until the end of the universe, or until the protons in the cores of the atoms decay, whichever comes sooner).

'Radioactive' simply means that the element is unstable, and will sooner or later decay into lighter elements.

About 90 of the elements (radioactive and non-radioactive) are found in nature, and the rest (radioactive) have to be made artificially.

Some of the radioactive elements (such as radium and uranium) take such a long time to decay that there are still some left — even though the universe is about 15 billion years old. But other radioactive elements decay fairly rapidly, so that there are none left in nature today. Plutonium is one of these. Practically all the plutonium on the Earth today was 'made' in a nuclear reactor. (However, you can actually find microscopic amounts of plutonium in uranium ores, as a result of the decay of heavier radioactive elements.)

Ancient History of Elements

Back in the early days, the thinkers tried to understand matter, and so they invented the idea of an 'element'.

The ancient Greeks thought that all matter was made up of just one element. But they disagreed about exactly what that 'element' was — Heraclitus thought it was 'fire', Thales suggested it was 'water', while Anaximenes said it was 'air'.

Another Greek philosopher, Empedocles, believed that all matter is made up from four elements — air, earth, fire and water.

At that time, seven of what we now call elements were known to the Greeks — but they didn't realise that they were elements! They were copper, gold, iron, lead, mercury, silver, and tin. The Greeks did get one thing correct — all elements have different properties.

NOBEL NAMES

The Chemistry Nobel Prizes for 1923 and 1966 were won, respectively, by Millikan and Mulliken. The Physics Nobel Prizes for 1954 and 1967 were won, respectively, by Bothe and Bethe. The Nobel Prizes for Physiology/Medicine in 1946 and 1948 were respectively won by Muller and Müller.

And who won the 1959 Nobel Prize for Literature? Quasimodo!

NAMES OF THE ELEMENTS

93 Neptunium, 1940–41, (after Neptune, the next planet out after Uranus).

94 Plutonium, 1940–41, (after Pluto, the next planet out after Neptune).

95 Americium, 1944–45, (after America).

96 Curium, 1944, (after Marie and Pierre Curie, the nuclear physicists).

97 Berkelium, 1949, (after Berkeley, home of the Lawrence Berkeley Laboratory, where it was first made).

98 Californium, 1950, (after California, and the University of California).

99 Einsteinium, 1952, (after Albert you-know-who. It was discovered in the radioactive debris of a hydrogen bomb).

100 Fermium, 1953, (after Enrico Fermi, the Italian-American Nobel Prize winning nuclear physicist. It was also discovered in the radioactive debris of a hydrogen bomb).

101 Mendeleevium, 1955, (after the Russian chemist, Dmitri Mendeleev).

102 Nobelium, 1957–58, (after Alfred Nobel, famous for the Nobel Prize).

103 Lawrencium, 1961, (after the American physicist, Ernest Lawrence).

104 Rutherfordium, 1969, (after Ernest Rutherford, the New Zealand physicist, who discovered the atom).

105 Dubnium, 1970, (after Dubna, where the Russian Joint Institute for Nuclear Research is).

106 Seaborgium, 1974, (after Glenn Seaborg, who was involved in the finding of 10 radioactive elements).

107 Bohrium, 1981, (after Niels Bohr, the Nobel Prize winner, who was a founder of quantum physics).

108 Hassium, 1984, (after Hassia, the Latin name for Darmstadt, where the GSI is located).

109 Meitnerium, 1982, (after Lise Meitner a German–Jewish female physicist, who many say deserved a Nobel Prize for her work).

Properties of Elements

Elements can be incredibly different from each other. Consider the property of 'Melting Point' — the temperature at which an element turns from a solid into a liquid. Boiling point ranges from –269°C for helium, to +3370°C for tungsten.

Take another property such as 'Density', which you can measure in grams per cubic centimetre. Densities range from 0.00008986 for hydrogen (the least dense) to 22.5 for iridium (the most dense). Other properties include how hard (diamond) or soft (lead) it is, how good it is at carrying heat or electricity (copper) and how much heat it can store.

ODD HEAVENLY NAMES

In 1916, the Russian astronomer Grigory Neujmin discovered a small asteroid between the orbits of Mars and Jupiter. He named it Gaspra after a Crimean health resort. On its way to Jupiter in 1991, the *Galileo* spacecraft zipped past Gaspra at close range. *Galileo* photographed many craters on this 15-kilometre-long rock. The craters were in turn named after famous spas and health resorts — Spa (Belgium), Bath (England), Aix (France), Baden-Baden (Germany) and Saratoga (New York).

Another heavenly body has a very special name. Because more than 200 Frank Zappa fans wrote to the International Astronomical Union, there is now an asteroid called (3834) Zappafrank.

Modern History of Elements

In the 1660s, the English chemist Robert Boyle grappled with the difficult concept of an 'element'. He thought that 'elements' could not be broken down to simpler substances. He didn't like Empedocles' 'elements' (earth, water, air and fire) for two reasons. Firstly, he couldn't extract or remove these 'elements' from other substances. Secondly, he couldn't combine these 'elements' to make other substances.

In 1789, the French chemist Antoine-Laurent Lavoisier released the first list of the elements. There were 23 of them! He also incorrectly included, as elements, a few very stable and hard-to-bust-apart chemicals (such as alumina and silica) which are not elements.

Beginning around 1808, Sir Humphrey Davy and Michael Faraday used the new technique of electrochemistry to split some chemicals. They showed that a few of these chemicals were not pure elements, but were made up of a combination of oxygen and another element. As more elements were discovered, the chemists tried to put them in some kind of understandable order.

By 1829, the German chemist Johann Wolfgang Döbereiner realised that some groups of three elements (such as chlorine, bromine, and iodine) shared similar properties. However, his fellow chemists didn't really appreciate the importance of his discovery.

In 1859, the German physicists Robert Wilhelm Bunsen and Gustav Robert Kirchhoff developed the spectroscope. Each element, when it is heated, will give off its own special coloured light. Way back in 1666, Isaac Newton had shown that a glass pyramid would split white light into its various colours. A spectroscope just used this principle to split the light from various heated elements into its unique patterns. More elements were discovered, thanks to the spectroscope.

In 1864, the British chemist John A. R. Newlands made a list of the elements, with the lightest ones first. He realised that every eighth element had similar properties. But this pattern worked only for the 18 lightest elements. Once again, other chemists didn't see the significance of this discovery.

Periodic Table of Elements

Finally, in 1869, the Russian chemist Dmitri Mendeleev proposed his first version of how to arrange the known elements in his Periodic Table of the Elements. The elements in each column all had similar properties. Suddenly, it became possible to look at gaps in the Periodic Table, and to predict the properties of elements in those gaps. For example, when Mendeleev set up his Periodic Table, he had to leave a vacant space between the elements calcium and titanium. Not only did he predict that an element would fit in there, he even predicted its properties. In 1879, scandium was discovered, and it had the predicted properties!

Around 1913–1914, the British physicist H. G. Moseley discovered a link between the atomic numbers of the elements and the frequencies of the X-rays that they could emit. For the first time, the chemists could give each element a unique atomic number. And for the first time, the chemists could work out exactly where in the Periodic Table of the Elements each element would go.

Newest Elements are Radioactive

With improved techniques, more and more elements have since been discovered. All of the elements heavier than polonium (atomic number 84) are radioactive. Some radioactive elements take billions of years to decay, while others do it in a few thousandths of a second. The short-lived radioactive elements have all decayed into smaller elements, so if we want them, we have to make them.

The artificial radioactive elements are usually made by smashing two lighter elements (such as lead and nickel) into each

LET THE ELEMENTS BE EVERYWHERE

The most abundant element in the universe is hydrogen (which has the atomic number of 1). It makes up about 90 per cent of all the atoms in the universe, and about 75 per cent of the mass of the universe. Helium (atomic number 2) is next. It makes up 7 per cent of all the atoms in the universe, and nearly 20 per cent of the mass of the universe. So the two lightest elements make up about 95 per cent of the mass of the universe.

However, you get quite different elements in the Earth's crust (the top 10–70 kilometres). The most abundant elements, in terms of how many atoms are present, are oxygen (46.6 per cent), silicon (27.7 per cent), aluminium (8.1 per cent), iron (5 per cent), and calcium (3.6 per cent).

element 114, in what they call a 'Nuclear Island of Stability'. Certainly, as we have worked our way up to element 112, the heavier elements have longer lives. Some physicists think that these more stable radioactive elements will have unusual properties that will be very useful. (Some people even claim UFOs are powered by elements from this island of stability!)

Naming of Elements

Along the way, the various committees had agreed on the names to be given to the elements up to element 103, Lawrencium. But in 1980, the International Union of Pure and Applied Chemistry (IUPAC) said that any elements heavier than atomic number 104 would not be named after mythological characters, people, or places. Instead they would be given names that would be the Latin equivalents of their atomic numbers. So element 104 would be called 'unnilquadium', element 105 would be 'unnilpentium', and so on.

By the late 1990s, there were another six unnamed elements to be introduced to the family of the elements. But not everybody was happy with giving them these new strange, Latin-sounding names.

The Name Game

other, and hoping that they stick together (it's a bit of a black art). The scientists do this using nuclear reactors, or particle accelerators. Unfortunately, the created element usually decays away incredibly rapidly. So you need scientists who are very skilled in experimental work to find this element in its very short life.

The scientists hope that they will find some long-lived elements up around

Glenn Seaborg, who shared the 1951 Nobel Prize in Chemistry, was strongly involved with the discovery of plutonium and nine other radioactive elements. His American colleagues (via the Committee on Nomenclature of the American Chemical Society, ACS) wanted to honour him by naming element 106 after him. Unfortunately, IUPAC had other ideas.

They did not want newly-discovered elements to be named after scientists who were still living, even though this happened with both Einsteinium (named after you-know-who) and Fermium (named after Enrico Fermi). They said that '*it is necessary to have the perspective of history in relation to these discoveries before such a decision is made*'.

The petty squabbling between these two chemists' organisations became ridiculous. Both the ACS and IUPAC put forth names for the six unnamed elements, and would not accept each others' suggestions.

Name Game Gets Worse

But then other groups of scientists got involved in this battle of the names.

There are three laboratories that have made all the 'new' elements. The Lawrence Berkeley Laboratory in California made elements 99 to 106. The GSI (Gesellschaft für Schwerionenforschung) Laboratory in Darmstadt in Germany has made all the elements from 107 up. But the Joint Institute for Nuclear Research at Dubna in Russia also claimed to have made elements 102 to 105. (It's actually quite tricky to be sure that you have a new element. When you are working with only three atoms, it's easy to miss an

---

LET THE ELEMENTS RUN FREE

About one-third of the elements exist freely (that is, not combined with other elements). These elements are not very active, so they tend not to react with other elements. They include copper, gold, nitrogen, platinum and the 'noble' gases (helium, neon, krypton, xenon and so on).

SEVEN ATOMS ARE ENOUGH

Seaborgium has a half-life of only 0.9 seconds.

What does half-life mean? If you have 100 atoms of Seaborgium, after 0.9 seconds, half of them (that is, 50) will have decayed into other elements, and 50 will be left. After another 0.9 seconds, half of the remaining 50 (that is, 25) will be left, and so on.

With such a short half-life, you have to be quick with your experiments. But in July 1997, scientists published a report that they had made seven atoms of Seaborgium (at the agonisingly slow rate of one atom per hour), and had successfully tested its chemistry.

It's amazing to think that you can do chemistry with only seven atoms, but the scientists did it.

They found that Seaborgium behaved like the other elements in its column in the periodic table — chromium, molybdenum and tungsten. At temperatures around 350°C, when it reacted with thionyl chloride gas, it made the same types of chemical compounds that tungsten and molybdenum did. And when Seaborgium reacted with liquid acids, it stayed neutral or made negative ions — again like tungsten and molybdenum.

---

atom or two. And sometimes, evidence for an element shows up only when you go back and analyse old data.)

The GSI laboratory put forward its own preferred names, but again, IUPAC disagreed. GSI then discovered elements 110 and 111 in 1994, and element 112 in 1996. So they protested against the IUPAC by refusing to nominate names for these elements.

This squabbling over names was eventually resolved in 1997, after a lot of back-room politicking. Finally, Glenn Seaborg got his name on an element, so we weren't left with a 'name vacuum'.

The fabulous thing about elements is that they're so pure and simple and uncomplicated. Thank goodness that the various groups of scientists resolved their differences. Otherwise, we might have ended up with long convoluted hyphenated names for something so elementary!

REFERENCES

'Getting a Place in the Periodic Table', *Science*, 14th October 1994, p 223.

Daleane C. Hoffman, '110, 111 . . . and Counting', *Nature*, Vol. 373, 9 February 1995, pp 471–472.

Dr Colin Keay, 'Astronomical Scam', *The Skeptic*, Vol. 15, No. 2, Winter 1995, p 52.

John. F. Kross, 'What's in a Name?', *Sky & Telescope*, May 1995, pp 28–33.

Ruth Lewin Sime, 'Lise Meitner and the Discovery of Nuclear Fission', *Scientific American*, January 1998, pp 58–63.

M. Schadel et al, 'Chemical Properties of Element 106 (Seaborgium)', Nature, Vol. 338, 3 July 1997, pp ix, 21, 22, 55–57.

Corinna Wu, 'Element 106 Takes a Seat at the Table', *Science News*, Vol. 152, 19 July 1997, p 44.

PLANETS LINE UP

If you watch enough bad television, or read sensationalist magazines, you will probably have heard that the Earth will be destroyed on the 5th of May, in the year 2000, when all the planets form a single straight line stretching out from the Sun. I've heard it claimed that the Earth will be tossed out of its orbit. At the very least, the gravitational stress on our fragile little planet will supposedly cause massive earthquakes and tsunamis, and billions of people will perish.

Is this true — will our little blue planet be damaged when all the nine planets from Mercury to Pluto make a straight line?

The End of Humanity?

The short answer is 'no'.

The planets won't line up! And even if they did, the combined pulls of the other planets would be smaller than the pull of a 747 cruising overhead!

To prove this, first we have to know more about how the planets are set up in our solar system — but along the way, we'll find a little mystery that even today, we cannot explain.

The Big Picture

So here's the big picture. Our Sun is one of the hundreds of billions of stars in the galaxy that we call the Milky Way. The nine known planets (Mercury, Venus, Earth, Mars, Jupiter, Saturn, Uranus, Neptune and finally Pluto) orbit around the Sun. All nine planets orbit in roughly the same plane. Mercury, the closest planet to the Sun, zips around in only 88 days, while distant Pluto takes about 247 years.

The early astronomers noticed that most of the points of light in the night sky stayed in the same position relative to each other. But a handful of them would wander across the sky in the course of a year — and that's how today we have the word 'planet' from the Greek word for 'wanderer'. The ancient astronomers could see five of the planets with the naked eye — Mercury, Venus, Mars, Jupiter and Saturn. For many years, astronomers thought that these were the only planets (besides Earth) to be found within our solar system.

Titius' Law – Bode's Law

But then, in 1766, the German scientist Johann Daniel Titius noticed that there was a strange mathematical relationship between the distance of each of the six known planets (six, if you include the Earth) from the Sun.

Now here come the numbers.

Imagine that we start off with the number 0, and then 3, and then we double the last number. So we end up with the numbers 0, 3, 6, 12, 24, 48, 96, 192, and so on. And then we add the number 4 to each number in our series, so now we end up with the new series 4, 7, 10, 16, 28, 52, 100, 196, and so on. And this is the great mystery. These numbers actually describe the relative

---

SOLAR SYSTEM OVERVIEW

Our solar system has one Sun, nine planets with some 63 satellites orbiting them, and a whole bunch of small bodies such as comets and asteroids.

The planets do not orbit in the plane of the equator of the Sun. Most of them are in a plane called the 'ecliptic', which is defined by the plane of the Earth's orbit. The ecliptic is inclined at seven degrees to the plane of the equator of the Sun. Pluto is inclined at 17 degrees to the ecliptic.

Most of the planets have circular orbits, although Mercury and Pluto have quite elliptical orbits.

ORIGIN OF SOLAR SYSTEMS

Over the years, the mathematicians have given us a fairly good model of how we think a solar system comes into existence.

It all begins with a big cloud of interstellar gas and/or dust. Something (such as the shockwave from a nearby supernova) disturbs it, and the cloud begins to collapse under its own gravity. As it collapses it gets warmer, and compresses in the centre. This first collapse supposedly takes less than 100 000 years.

The new centre now has enough mass so that the rest of the gas and dust orbits around it. Most of the gas flows into the forming star, and makes it heavier. But because the star-to-be is rotating, it has an outward centrifugal force that stops some of the gas actually reaching the forming star. This gas makes a disc around the forming star, which is called the 'accretion disc'.

The accretion disc is some distance away from the warm centre, and so it cools off. First the metals condense, and then the rocks condense. The particles of dust in the accretion disc collide and form larger particles — as big as cars, buildings or small mountains. Once a particle turns into a mountain, it has enough gravity to suck in other particles, so it begins to enter a period of runaway growth. These large objects are called planetismals. Planetismals collide with each other and gradually form planets.

About ten to 100 million years after the process began, you end up with half-a-dozen to a dozen planets orbiting around the central star.

distances of each of the planets from the Sun! So Mercury is four units out, Venus is seven units out, Earth is 10 units, and so on.

At the time, no one paid much attention to this amazing coincidence, until 1772, when the astronomer Johann Elert Bode included it in an introductory textbook. Because he made this mathematical relationship popular, it became known as Bode's Law. Perhaps, thought the other astronomers, this equation could be used to find other planets, even further away from the Sun.

BODE'S LAW			
DOUBLING	ADD 4	PLANET	RELATIVE DISTANCE
0	4	Mercury	3.9
3	7	Venus	7.2
6	10	Earth	10
12	16	Mars	15.2
24	28	Asteroid Belt	21.6–31.8
48	52	Jupiter	52.0
96	100	Saturn	95.56
192	196	Uranus	191.9
384	387	Neptune	301.1
768	772	Pluto	395.3

SOLAR SYSTEM RECORDS

In our solar system there are 17 bodies which are bigger than 2000 kilometres across. Beginning with the biggest, they are the Sun, Jupiter, Saturn, Uranus, Neptune, Earth, Venus, Mars, Ganymede, Titan, Mercury, Callisto, Io, our Moon, Europa, Triton and Pluto.

There are 13 moons which are less than 40 kilometres across. They are Deimos and Phobos (Mars), Leda, Adrastea, Ananke and Sinope (Jupiter), Pan, Calypso, Atlas, Telesto and Helene (Saturn), and Cordelia and Ophelia (Uranus).

Saturn is less dense than water. So if you had a big enough ocean, it would float.

Bode's Law Finds Planets

When the seventh planet, Uranus, was discovered at the distance predicted by Bode's Law, astronomers began to suspect that even more planets could be out there, waiting to be discovered.

Bode himself pointed out that according to his law, there seemed to be a planet missing, somewhere between Mars and Jupiter. The gap between these two planets was much bigger than the law predicted. When the astronomers went searching in the approximate location of this 'missing planet' they found, not an extra planet, but the giant

SIZE OF SOLAR SYSTEM

The diameters of the Earth, Jupiter and the Sun are roughly in the ratios of 1 to 10 to 100.

Imagine that we shrink the solar system one billion times. Then the Sun is around 1.5 metres (5 feet) in diameter — roughly the height of a smallish teenager. Then the Earth would be about 150 metres (about 164 yards) away from the Sun (one city block) and about 1.3 cm (half an inch) in diameter (the size of a grape). The Moon is about 30 cm (one foot) away from the Earth. Jupiter is about 15 cm (6 in) in diameter (the size of a large grapefruit), and about five blocks from the Sun. Saturn is twice as far away again (10 blocks), and a bit smaller (the size of an orange). Uranus and Neptune would be the size of lemons, and about 20 and 30 blocks from the Sun respectively.

On this scale, a human would be the size of an atom, and the nearest star would be over 40 000 kilometres (25 000 miles, or the circumference of the Earth) away.

asteroid, Ceres. The astronomers then rapidly discovered a whole bunch of other asteroids. The discovery of this asteroid belt (between Mars and Jupiter) confirmed Bode's Law.

And so the hunt was on for the eighth planet. But when *it* was finally discovered, Neptune did *not* lie at the distance predicted by the law — it was much too close to the Sun. So for a while, Bode's Law was discredited, because it was so hopelessly inaccurate. But when Pluto, the ninth planet was discovered, it was close to where Neptune should have been. Perhaps Pluto was an interloper from deep space that shoved Neptune into a closer orbit, leaving Pluto in Neptune's original orbit, or perhaps not . . .

But anyway, Bode's Law worked very well for the first seven planets and the asteroid belt. And it did help astronomers discover the planet Uranus. Even to this day we have no explanation for why Bode's Law gave the positions of these planets so accurately.

The Planets Go Around and Around

But while we don't know why Bode's Law predicted the positions of the planets, we *do* know a lot more about the claimed lining up of the planets that will supposedly happen on May the 5th in the year 2000. It is ridiculously easy to predict just where the planets will be in a few years' time (it's harder when you try to predict a million years into the future, but that's another story). It's also easy to see where the planets were in the past. In fact, there are several sites on the web where you can plug in the date of your birth, and see exactly where the planets were when you were born. (One such site is at http://www.fourmilab.ch/cgi-bin/uncgi/Solar).

The nine planets are all whizzing around the Sun, in roughly the same plane. Time is just God's way of making sure that everything doesn't happen at once, and so, if you give the planets enough time, eventually, some of the planets will line up. It turns out that a line-up of the kind that will happen in 2000 is actually fairly common. In fact, it happens about every 20 years. But the line is never perfectly straight. Even during what astronomers call a 'planetary alignment', the line is still very jagged.

20-YEAR ALIGNMENT

Why 20 years?

The reason is that Jupiter takes about 11.9 years to make a complete orbit of the Sun. Saturn takes about 29.7 years for a complete orbit. By the time that Jupiter has made a complete orbit, Saturn has moved on. By the time that Jupiter has caught up with Saturn, so that there is a straight line joining the Sun, Jupiter and Saturn, 20 years have passed. Jupiter and Saturn line up with the Sun every time that Jupiter has done 1.7 orbits, and Saturn has done 0.7 orbits.

The 2000 AD alignment is not particularly close. The one in 2020 is a little tighter — most of the planets will lie within a slice (with the pointy bit on the Sun) about 12 degrees wide.

Doomsday - 5th May, 2000 AD

So what will happen on the supposed Day of Doom — the 5th of May, 2000 AD?

The five planets which we can see without a telescope (Mercury, Venus, Mars, Jupiter and Saturn) will be in a very jagged line on one side of the Sun. How jagged is this jagged line?

Alignment - 1° or 180°?

Think of how you cut a birthday cake. If you cut a cake into 14 wedges, each will be about 25 degrees across. In our solar system, in the year 2000, Mercury, Venus, Mars, Jupiter and Saturn will be scattered across 25 degrees — not 1 or 2 degrees. Uranus and Neptune will actually be about 90 degrees off to one side. Pluto will be about 160 degrees away from

ASTEROIDS

The first asteroid was discovered on the first day of the 19th century. On the 1st of January, 1801, Giuseppe Piazzi discovered a strange object in the sky. He called it Ceres, after the Sicilian Goddess of Grain. By the end of the 19th century, several hundred asteroids had been discovered, and we now know of 7000.

The combined mass of all of the asteroids known so far is less than the mass of our Moon. Ceres is the largest asteroid discovered. It is 933 kilometres across, and accounts for about one quarter of the mass of all of the asteroids so far discovered.

Twenty-six of the asteroids are over 200 kilometres across. We have probably found 99 per cent of the ones that are bigger than 100 kilometres across. We have probably found half of the ones that are in the range of 10 to 100 kilometres across. We don't really know much about the smaller ones. There might be a million that are one kilometre across.

There are three main types of asteroids — Main Belt Asteroids, Near-Earth Asteroids, and Trojans.

Main Belt Asteroids spend their time between Mars and Jupiter.

Near-Earth Asteroids will occasionally come close to the Earth. The amount of money so far spent looking for Near-Earth Asteroids ($US25 million) is half the amount spent on making the movie *Deep Impact*, which tells the story of an asteroid hitting the Earth.

The Trojans are located near gravitationally 'stable' points in Jupiter's orbit — 60 degrees ahead of Jupiter, and 60 degrees behind Jupiter. These points are called Lagrange Points. There are several hundred Trojans known. We don't know why, but there are more Trojans ahead of Jupiter than behind it.

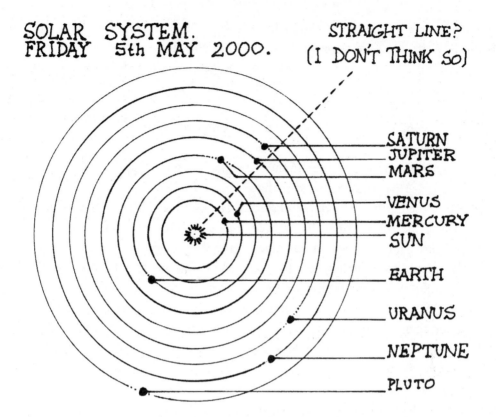

SOLAR SYSTEM.
FRIDAY 5th MAY 2000.

STRAIGHT LINE?
(I DON'T THINK SO)

SATURN
JUPITER
MARS
VENUS
MERCURY
SUN
EARTH
URANUS
NEPTUNE
PLUTO

NOT TO SCALE: FOR CLEARER VIEWING, SYSTEM
IS SHOWN BUNCHED UP.

Mercury, Venus, Mars, Jupiter and Saturn. And Earth will be another 20 degrees further around. In fact, Earth will be on the other side of the Sun away from Mercury, Venus, Mars, Jupiter and Saturn!

When I first heard about the Great Planetary Alignment of the year 2000, I thought that *all* the planets would be in a sector maybe 1 or 2 degrees wide. Having five of the planets scattered over 25 degrees, and the remaining four planets over another 160 degrees suddenly doesn't sound very impressive at all.

So one big problem with the big Line Up of the Planets in 2000 AD is that the planets won't be lined up at all!

But the big question is, what will happen to our planet on May 5 in the year 2000?

Will the alignment of the planets be close enough to cause damage? Of course, none of us can predict the future, but we can learn from what has happened in the past.

The 1982 Grand Planetary Alignment

Back in 1982, an even better alignment of the planets happened. All nine of the planets were on the same side of the Sun, scattered over some 90 degrees, in what the astronomers call a 'Grand Alignment'. (In fact, this alignment allowed the *Voyager 1* and *2* spacecrafts to leapfrog their way to the outer planets, using the manouevre called a 'Gravitational Slingshot'.) And just like this time, people in 1982 were worried that it would have devastating effects on life as we know it.

In 1974, the astrophysicist John Gribbin predicted in his book, *The Jupiter Effect*, that when the planets lined up, the Earth would be caught in the middle of a huge gravity struggle between the Sun and the planets, especially the giant planet, Jupiter.

These stresses, he claimed, would change the speed of the Earth in its orbit, and shift the centre of the solar system. He warned that geological fault lines would be ripped open, causing massive earthquakes. In fact, he went so far as to predict that 'in 1982, when the Moon is in the seventh house, and Jupiter aligns with Mars and with the other seven planets of the solar system, *Los Angeles will be destroyed*'!

Well, as predicted, the grand alignment happened, with the planets scattered over some 90 degrees (which is nothing like a straight line). For better or worse, Los Angeles was not destroyed, and life as we know it went on. In fact, the amount of earthquake activity that year was completely normal.

Why It Doesn't Happen

The problem with Gribbin's theory was that the effect Jupiter's gravitational force has on the Earth is tiny compared to the Sun's. So there was no gravity struggle at all. And anyway, since the Moon is so much closer to us, its gravity pull is stronger than all the other planets put together.

A few months after the grand alignment happened, Gribbin admitted that his predictions were wrong. Yes, he said, he *had* made a mistake. So late in 1982, Gribbin

GRAVITATIONAL EFFECTS OF THE SOLAR SYSTEM ON EARTH

BODY	GRAVITATIONAL PULL
Sun	330 000
Mercury	0.15
Venus	10.6
Earth	-
Moon	1680.0
Mars	0.39
Jupiter	18.0
Saturn	1.31
Uranus	0.044
Neptune	0.020
Pluto	0.000002

As you can see, the Sun (330 000) has the greatest gravitational effect (or pull) on the Earth. The Moon (1680) follows a long way behind. The Moon and the Sun are each stable in their distances away from the Earth, so their gravitational effects don't change much.

The next major gravitational effects are those of Jupiter (18) and Venus (10.6) — about 100 times weaker than that of the Moon. Jupiter is a long way from the Earth, but it is very heavy — over 90 per cent of the mass of all the planets and moons of the solar system is in Jupiter. Venus is not very massive, but it is quite close, so it has a relatively high gravitational effect on the Earth.

If you ignore the Sun, the Moon has the main gravitational effect on the Earth. Jupiter and Venus have an effect about 100 times weaker. The other planets are much weaker again.

changed his mind. He now claimed that the real danger would be when the Earth was on the *other* side of the Sun, away from the other planets — on 2nd November, 1982. The pull of all the planets *plus* the Sun would definitely have an effect — not just on the geology of the Earth, but on the weather as well. In fact, he warned people to get ready for freezing winters, and perhaps even a little ice age.

Gribbin did have some research to back up his claims. A Chinese study had linked very cold winters in the northern hemisphere to planetary alignments. Unfortunately, this research was actually based on Gribbin's own earlier predictions, including the idea that a line up of the planets would change the Earth's orbit. The second predicted catastrophe didn't happen either.

Recently, some of Gribbin's predictions have been dragged back out again, and various sensationalist magazines and TV shows are warning that we are all doomed.

The Forces are Tiny

Now if you do the numbers, the *gravitational pull* of the five planets on the other side of the Sun is about 30 millionths of the gravitational pull of the Sun — pretty close to nothing!

But let's forget the gravitational pull, and look at the *tidal forces* caused by these planets. The tidal force is how another astronomical body (such as our Moon, or the Sun, or another planet) can pull more strongly on one side of our planet than on the other. It turns out that for the 5th of May in the year 2000, the tidal pull of a jumbo jet flying at 30 000 feet is greater than the tidal pull of all the five planets lined up together!

And if that doesn't convince you, back on 31 January in 1962, the same five planets were lined up better than they will be in the year 2000, and nothing happened.

So the apocalyptic line dance of the planets goes on . . . but it's not really all that apocalyptic after all! And like all line dances, it just goes on and on and on . . .

REFERENCES

Isaac Asimov, *Asimov's Biographical Encyclopedia of Science and Technology*, Pan Books, 1975, pp 208–209.

John Gribbin, 'Stand by for Bad Winters', *New Scientist*, No. 1329, 28 October 1982, pp 220–223.

C.L. Stong, 'Graphs That Predict When Planets Will Line Up With Another Planet or the Sun', *Scientific American*, August 1975, pp 116–120.

Richard Talcott, 'Ask Astro', *Astronomy*, October 1997, p 86.

http://www.fourmilab.ch/cgi-bin/uncgi/Solar

TV, HEART ATTACKS AND CPR

Television is an important source of information for many people, and for some stories TV is great — moving pictures are fantastic for wildlife documentaries. But sometimes television is misleading and, in some cases, this can be very important. If TV presents medical knowledge, for instance, and it's inaccurate — people may believe it and make foolish choices about their health.

Communication

For most of our history, a message could travel only as fast as a human could run, a horse could gallop, or a pigeon could fly.

People sometimes used fires to send urgent messages. The Greeks once sent a message (from Troy to Argos in Greece) some 800 kilometres (500 miles) in a few hours, via a series of bonfires. The message told the folks back home that Troy had finally fallen. Similarly, in 1588, the English sent a message from Plymouth to London (300 kilometres in 20 minutes) via the 'fires-on-hilltops' method, to warn of the approach of the Spanish Armada.

But for all regular communications, people were limited to the speed of some kind of animal.

Electricity is Used

The 'communications revolution' began in 1833, when Carl Friedrich Gauss and Eduard Webber sent electrical pulses down a copper wire. By 1838, the first electric public telegraph was operating in the United Kingdom, while in the USA, Samuel Morse had invented Morse Code. He also invented the relay, which allowed signals to be sent greater distances. By 1851, the electric telegraph had spread via underwater cables across the English Channel. By 1872, it had reached South Australia.

Key Discovery for TV

No one single person 'invented' television. Instead, TV came into existence gradually as many people in different countries made separate inventions and discoveries.

In 1873, Louis Joseph May, an English telegraphist, discovered that when light shone on the element selenium, it reduced the resistance of the selenium. He designed (but did not build) a system in which an image was broadcast onto a screen made up of thousands of selenium cells. Each selenium cell saw only a small part of the image. The selenium cells were electrically connected to light bulbs (some distance away) via batteries. So if bright light shone on one selenium cell, the resistance of that selenium cell would drop, and allow a lot of electricity to flow from the battery into the remote light bulb, which would shine brightly.

The discovery of selenium's strange property meant that, for the first time, a human-built device could 'read' the brightness of light — and that a remote set of light bulbs could 'show' what had been read. This was one of the key concepts needed for television.

Scanning the Image

In 1875, George Carey from Boston (USA) worked out how to send TV images. He wanted to break the image into many

TELEVISION IN SPACE

In 1965, the first of the Mariner series of spacecraft sent back photographs of Mars. In July 1969, TV cameras sent back live colour broadcasts of the first landing on the Moon.

separate picture elements (just like today's system), but to send each picture element simultaneously, over a separate circuit. Today, we call each picture element a 'pixel'.

In 1881, W. E. Sawyer (USA) and Maurice Leblanc (France) came up with a better idea. They proposed systematically scanning the image, once it came out of the lens. The image was to be read one line at a time, beginning at the top and then working slowly down to the bottom.

Nipkow's Scanning Disc

On 6 January 1884, Paul Gottlieb Nipkow (Germany) patented a mechanical rotating disc which would allow this scanning. He also sketched out the system. This simple system was the basis of the CBS colour system, released in 1951. It was the best mechanical system ever built. But it could never be as good as an all-electronic colour system, with no moving parts.

HOW A TV WORKS

If you want to invent a television system, you have to solve four major problems.

First, you have to scan the image into many tiny dots or pixels. Usually, these pixels each have a different brightness. You then have to convert these pixels into an electrical signal. The key discovery here (in 1873) was that light falling on selenium changed its resistance.

The second problem is to transmit this electrical signal through the air. In 1901, Marconi had sent radio signals across the Atlantic. No fundamentally new technology had to be developed.

The third problem is to change the electrical signal back into dots of light in the television receiver in the home. That was done with a cathode ray tube (Braun, 1897).

The fourth problem is to show these individual pictures (or frames) rapidly enough to fool the human eye. This was tricky. The Nipkow scanning disc (1884) solved this problem in a mechanical way. Compared to electronic systems, it was limited in its top speed and accuracy; but in the early days, the Nipkow system was easier to build and get running.

Luckily, there is a phenomenon called 'persistence of vision', that helps solve this fourth problem. If five individual images (frames) are shown in one second, they will appear, to the human eye, as flickering pictures. But if 25 to 30 individual frames are shown in one second, they will appear, to the human eye, as a perfectly smooth moving picture — each frame will 'persist' and blur seamlessly into the next frame. (Actually, ten frames/second will usually appear smooth and free of flicker, but showing 25 to 30 frames/second gives a big margin of safety.)

A Monitor to See the Image

In 1897, Karl Ferdinand Braun (Germany) built a cathode ray tube (CRT). For most of the history of television, the CRT has given us the picture. But there were two problems with his early CRT. First, the beam was not very bright. But in 1898, A. Wehmelt (Germany) invented an alkaline oxide cathode which was much brighter. Secondly, the focusing was too coarse which produced a blurry image. But in 1899, Vichert (France) invented a concentration electrode, which gave a much sharper image.

The CRT uses high voltage (10 000–20 000 volts) to accelerate electrons towards the screen. A chemical called a *phosphor* is coated on the inner surface of the glass tube. The phosphor (often oxides of zinc and sulphur) will glow with a bluish white light when the high-energy electrons hit it. The more energy that you have in the electrons, the brighter the TV picture will be.

In 1907 Professor Boris Rosing (Russia) used a rotating mirror drum to scan the image, and the CRT for showing the image. He was able to transmit crude geometrical shapes, but no halftones. (He 'disappeared' in 1918, during the Russian Revolution).

A Theoretical Visionary

On 18 June 1908, A.A. Campbell Swinton published a remarkable letter in Nature entitled 'Distant Electric Vision'. He proposed using CRTs in both the TV camera and the TV receiver, and he outlined the

TV RESOLUTION

The more lines there are in a frame, the sharper the image. Baird and his early fellow inventors started off with 10 to 30 lines.

On commercial TV, the UK used 405 horizontal lines in each frame until 20 April 1964, when it changed to 625 lines. The USA, Japan and Latin America use 525 lines, while Australia and much of Europe uses 626 lines. In 1995, High Definition TV sets, capable of receiving 1125 lines, first went on sale in Japan — but the majority of sets still run on 525 lines. France and Monaco enjoy the sharpest analogue image of all — 819 lines.

A lot of pixels get broadcast. Consider a typical US broadcast. Thirty frames are sent out each second, and each frame has 525 lines. There are about 330 pixels in each line, so there are about 170 000 pixels in each frame. In one second, about 5 million pixels get transmitted.

ODD TV FIRSTS

- Michel Lotito of Grenoble in France (Monsieur *Mangetout* = Eat All) eats about one kilogram of metal each day. Besides eating ten bicycles, a coffin and a Cessna light aircraft (!), he has also eaten seven TV sets.
- The tallest human-built structure was the Warszawa Radio and TV mast, 100 kilometres northwest of Warsaw. It weighed 617 tonnes (606 tons), and was finished on 18 July 1974. It was 646.4 metres (2120 ft 8 inches) tall, but it fell down during renovation work, on the 10 August 1991. Since then, the Poles have called it the 'World's Longest Tower'.
- The highest circulation of any weekly periodical was reached in 1974, by the *TV Guide* in the USA. It was the first magazine to sell more than a billion copies in one year.
- The largest TV set ever built was the Sony Jumbotron, at the 1985 Tsukuba International Exposition in Japan. It measured 24.4 by 45.7 metres (80 ft x 150 ft). I had my face on it for about four minutes in 1985, while I was reporting on it for *Quantum*.
- The longest pre-scheduled telecast ran for 163 hours and 18 minutes. From the 19th to the 26th of July 1969, GTV 9 in Melbourne in Australia covered the *Apollo 11* moon landing.
- Vanna White, the hostess of the American TV show *Wheel of Fortune*, claps her hands 720 times per show, or around 28 000 times per year.

fundamentals of a complete TV transmission system. He referred to his system as 'an idea only', because the technology to build it was not available at that time.

In 1919, Denes von Mihaly (Hungary) transmitted pictures of instruments and moving letters over a distance of a few kilometres.

On 29 December 1923, Vladimir Kuzmich (Kosma) Zworykin (a student of Rosing) applied for a patent on a camera tube which used a stored-charge image (iconoscope). This proved that the theories of A.A. Campbell Swinton were valid. (There was a delay in issuing his patent — it was not issued until 20 December 1938.)

Baird – Successful Inventor, Commercial Failure

In February 1924, John Logie Baird (UK) transmitted the image of a Maltese Cross over a distance of some three metres (10 feet) using Nipkow's mechanical scanning system.

From 1 to 27 April 1925, Baird gave the first public demonstrations of TV from the first floor of Selfridges store. On 30 October 1925, he transmitted the first image of a human (a 15-year-old boy, William Taynton). And, on 27 January 1926, Baird and Charles F. Jenkins (USA) electrically transmitted halftone images. Each frame

TV MAKES YOU FAT

There does seem to be a modern epidemic of obesity. In Australia, about one-fifth of everybody aged up to 25 years of age is obese. This rises to 60 per cent of the population when you look at people over the age of 45.

One American study, looking at 746 children aged between 10 and 15, found that children who spent more than five hours a day watching TV had a 4 to 5 times increased risk of being overweight. According to their statistics, there was a 60 per cent association between being obese, and watching lots of TV.

Another study looked at the metabolic rate of a group of 8- to 12-year-old children. They found that the basal metabolic rate (how much energy they burn) decreased in the children when they looked at TV. This study measured the children only for brief periods.

However, another study by Maciej Buchowski (at the Meharry Medical College at Vanderbilt University in Nashville) found that watching television uses up 20 per cent more energy than simply lying down and doing nothing. This study followed the children for 24 continuous hours at a time.

So these two studies are very contradictory, but there is one thing we do know. Over the last ten years, the number of advertisements promoting high-sugar high-fat foods have increased enormously in the children's viewing TV hours.

(or image) had only 30 lines of resolution, and was refreshed 10 times each second. ('Refreshed' means that a new frame was transmitted.)

The BBC (British Broadcasting Corporation) was created in 1927 by Royal Charter. Right from the beginning, its aim was to 'educate and enlighten', and entertain as well. This is quite different from the American system, where the commercial TV stations try to find out what the viewing public wants, and then gives it to them.

In 1927, the Bell Telephone company transmitted images between New York and Washington, D.C. It used the system designed by May back in 1873. At the transmitter, it used 2500 selenium cells to pick up the image. At the receiver, it used 2500 neon lights to show the image.

Colour TV in 1928

On 3 July 1928, Baird demonstrated colour TV, with 30-line resolution. He worked with the three primary colours — red, green and blue. The image was scanned by a spinning Nipkow disc, which had three separate spirals of holes, one for each colour. At the receiver, he used mercury vapour to give green, helium gas to give blue, and neon gas for red.

In 1929, H.E. Ives and his colleagues from the Bell Telephone Company used the Nipkow disc to transmit colour images from New York City to Washington, D.C., — but at the very low resolution of only 50 lines. They used three separate channels to carry three separate primary colour images, which were blended in Washington into a single low-resolution colour image.

On 30 September 1929, Baird transmitted TV on an unused BBC radio channel for 15 minutes. Thirty TV sets were able to pick up the signal.

In May 1930, he sold the first TV set — the Baird Televisor.

Pioneering TV Broadcasts

On 22 August 1932, the BBC had its first pioneering TV transmissions. The 30-minute TV signal went out every night (except Thursday) after 11 pm (when the radio broadcasts had finished for the evening). Also in that year, the Radio Corporation of America (RCA) demonstrated their all-electronic (no spinning disc) TV system, using 331 lines. It used Zworykin's iconoscope as a camera tube, and Braun's cathode ray tube in the TV receiver.

In 1933, the British public could buy the 'Silvatone Souvenir'. This was a home

FAMOUS TV QUOTES

John Logie Baird found a lot of resistance when he tried to get people interested in television. He said in 1940, 'They were convinced that the transmission of images — especially mentioning fog as an impediment — was impossible.'

In 1936, Rex Lambert, a radio critic in London, wrote, 'Television won't matter in your lifetime or mine'.

In 1948, Mary Sommerville, a pioneer of radio, said, 'Television won't last. It's a flash in the pan.'

Orson Welles said in the *New York Herald Tribune* (12 October 1956), 'I hate television. I hate it as much as peanuts. But I can't stop eating peanuts.'

Alfred Hitchcock was quoted in *The Observer* (London, 19 December 1965) as saying, 'Television has brought back murder into the home — where it belongs.'

Groucho Marx was quoted in Leslie Halliwell's *Filmgoer's Companion* (1984) as saying, 'I find television very educational. Every time someone switches it on I go into another room and read a good book.'

In London, on 5 September 1990, Ed Turner was quoted in *The Daily Telegraph* with these words, 'If we had had the right technology back then, you would have seen Eva Braun on the Donahue show and Adolf Hitler on *Meet the Press*.'

FIRST CABLE AND COLOUR TV IN OZ

In 1973, Bush Video set up the first cable TV network in Australia. They did this at the Aquarius Festival at Nimbin (NSW). They used this network to broadcast the first colour TV signals in Australia.

Michael Glasheen was the guiding force behind Bush Video, and I was one of the many helpers. I bought one kilometre of coaxial cable in Sydney, loaded it into the boot of my 1954 Chevrolet (which immediately flattened its rear springs) and drove some 800 kilometres (500 miles) to Nimbin.

For a week or two, we had two-way live cable TV stretching over a kilometre or so, with a central studio, and several remote monitors and cameras.

recording kit, which could record radio, or even TV. The signal was recorded with a steel needle that cut into a soft aluminium disc, spinning at 78 rpm. The kit came with six aluminium discs. Four minutes of a BBC TV transmission in April 1933, had been recorded on one of the discs. There were four frames transmitted each second, and only 30 lines on each frame. This gave a tiny bandwidth — about nine hundred times smaller than today's TV bandwidth. (In 1996, an English enthusiast, Don McLean, successfully used modern technology to get a recognisable signal from this disc. He had to clean the badly-corroded disc, and compensate electronically for the fact that the recording motor — back in 1933 — did not spin at a perfectly even rate. Finally, he was able to see pictures of high-kicking dancing girls.)

In March 1935, Germany began transmitting three days a week at 180 lines. Soon they had the world's first TV announcer, Ursula Patzsche. (They kept on transmitting until mid-1943, when the studio was destroyed by Allied bombing.)

The 1936 Olympic Games at Munich were the first to be 'televised'. However, the images were not broadcast through the air, but sent via cable to 28 'public television parlours' (Fernsehstuben) in Berlin that showed up to eight hours of live broadcasts from the games.

'Official' TV

The first 'official' regular BBC public television transmissions began on 2 November 1936, from Alexandra Palace. At that time, there were only about 100 TV sets in the UK. The Baird (240 lines) and the Marconi-EMI (405 lines) systems broadcast on alternate weeks. The Baird system was dropped on 4 February 1937, because its camera gave an inferior picture, and was so noisy that it had to be bolted to the floor inside a soundproof cubicle.

In the USA in 1936, the RCA transmitter on the top of the Empire State Building transmitted the first commercial TV (at 343 lines) in the USA, to an audience of some 5000 viewers.

On 4 February 1938, Baird showed colour TV images (at 120 lines) to an audience of 3000 amazed people. The images travelled through the air some 13 kilometres (8 miles) from his studio at Crystal Palace to the Dominion Theatre in London. The image appeared on a large screen, 4 m x 3 m (13 ft x 10 ft).

The 'official' launch of American commercial TV was on 30 April 1939, at the opening of the New York World's Fair in Flushing Meadows. The first images showed the huge symbols of the fair — the Trylon and Perisphere. Then the cameras moved around to show the sights of the fair — the parades, the crowds, the celebrities (such as President Franklin D. Roosevelt) and the foreign visitors in their exotic national costumes. The images were relayed to the top of the Empire State Building, and then to a big screen at the Radio City Music Hall as well as to TV sets in department store windows. People began to buy TV sets.

Also by 1939, Germany had made the world's highest definition (at 1029 lines) colour TV sets (built by Telefunken). In 1939, the BBC in England had an audience of some 20 000 TV sets.

World War II Interrupts Mickey Mouse

On 1 September 1939, as World War II began, the BBC suddenly stopped transmitting in the middle of a Mickey Mouse cartoon. They were worried that the German bombers might home in on the transmitted signal. (But on 7 June 1946, BBC TV resumed with the same Mickey Mouse cartoon. Jasmine Bligh said, 'Hello, do you remember me?' Leslie Mitchell followed with, 'As I was saying when we were so rudely interrupted . . .')

The first regularly operating TV station in America was WNBT. It went on air in 1940. The signal was received by around 10 000 TV sets, most of which belonged to RCA (which also owned WNBT) employees.

TV Lies Well – Experiments Reveal!

Soon TV spread throughout wealthy countries, and then into poorer countries. But as it went, it carried both truth and lies.

In fact, research suggests that viewers find it more difficult to tell the difference between a lie and the truth with television, than they do for radio or newspapers.

This is quite surprising considering that with newspapers we have only the raw text of the words. On radio we get more than just the text — we get various vocal cues (change in pitch of the voice, hesitations, pauses, and so on). And on TV we get all three — the text, the vocal cues and the visual cues (body moves, facial expressions and eye contact). So you would think that the extra information would help us pick a liar — but it seems not.

TELEVISION AND VIOLENCE

According to the American Psychological Association, by the time he or she is 11 years old, the average American child has seen on TV some 8000 murders, and 100 000 lesser acts of violence and brutality. Does this affect children? To start answering this question, you need to find a place where television has just been introduced, and then you need to trace the children's lives carefully.

Dr Tony Charlton, of the Cheltenham and Gloucester College of Higher Education, is studying children who live on St Helena — an island that didn't have television until March 1995.

St Helena is a remote island — 3200 kilometres east of South America, and 1600 kilometres away from western Africa. It has a population of 5500, of whom 1300 are schoolkids ranging from preschool to secondary school.

Tony Charlton looked at these pupils *before* television arrived. He found that they had one of the lowest incidences of behavioural problems anywhere in the world (3.4 per cent).

There's a lot of violence on TV, on shows, in the news, and even in commercials. Dr Charles Anderson from Hennepin County Medical Center, in Minneapolis, examined the TV commercials shown in America during the 1996 Major League Baseball Play-Off.

He reviewed 15 televised games and counted 1528 commercials. 104 (6.8 per cent) of the commercials contained violent content. 'In these 104 violent commercials, 69 contained at least one violent act, 90 contained at least one violent threat, and 27 contained evidence of at least one violent consequence.'

There were burning bodies and bloody corpses, forceful restraint and stabbings, as well as shootings and guns held to victims' heads. Dr Anderson writes that, 'the effects of violence in commercials are likely to be negative for society for the following reasons: (1) perpetrators of violence will likely go unpunished, (2) the short term negative consequences of violence for the victim are blunted as they may not appear to have caused harm or pain, and (3) long-term repercussions such as emotional, financial, social and physical disabilities are extremely *unlikely* to be expressed in a commercial.'

A 1972 report by the US Surgeon General and National Institute of Mental Health said that kids who watched TV violence were made more aggressive.

It will take several more years before Dr Charlton's results about St Helena are in. He has published *one* paper so far that looks at the behaviour problems in the preschool children now that TV has entered their lives — problems such as poor concentration, temper tantrums, lack of sociability with their peers, and fighting or destructive behaviour. It seems as though there has been an increase in behavioural problems with the arrival of television.

Sir Robin Day, a British political commentator, did an unusual experiment with Dr Richard Wiseman, a psychologist, and his colleagues at the University of Hertfordshire in the UK. The subjects were some 41 000 newspaper readers, radio listeners and TV viewers.

Sir Robin Day was interviewed twice on a fairly neutral topic — his favourite films. He lied all the way through one interview, and told the truth all the way through the other.

Both television interviews were shown on the *BBC Tomorrow's World* program. The radio versions were broadcast on *BBC Radio 1*. The transcripts (the raw text) of these two interviews were printed in *The Daily Telegraph*. People were asked to decide which interview told the truth and which contained the lies, and to phone in with their answer.

CHARLES P. SCOTT

**Maybe we should leave the last word to Charles Prestwich Scott, the editor of *The Manchester Guardian* from 1872 to 1929, who didn't like the name 'television' for this newly invented device.
'*Television?* no good will come of this device. The word is half Greek and half Latin.'**

TV Lies – The Results

According to Richard Wiseman's paper in *Nature*, the radio listeners picked the lies 73.4 per cent of the time, newspaper readers 64.2 per cent, and television viewers 51.8 per cent. Fifty-one point eight per cent is virtually the same as chance, which is 50 per cent! So, radio listeners can pick a lie three-quarters of the time.

One problem with this study is that there was only one single subject (Sir Robin Day), not a selection of subjects. Another problem is that the audience was self-selecting (they decided if they would phone in with an answer, or not). Perhaps the people who can easily pick lies decided not to join in this study. Yet another problem is that the audience could have lied when they rang in!

But if you can believe this study, the lesson is clear — if you want to tell lies, and get away with it, tell them on TV.

One of the lies that TV medical shows have spread is that Cardiopulmonary Resuscitation (CPR) will work well on practically everybody.

Colour - The Bitter Fight

Colour TV was born from a vicious corporate fight between CBS and RCA. The CBS and RCA engineers had each built quite different (and non-compatible) colour TV systems. Only one system could be accepted by the NTSC (National Television Standards Committee). The winner of course would become very wealthy.

CBS Mechanical Colour System

The CBS tricolour system sent out three separate colour images — one red, one green and one blue.

There were two major problems with this. First, the signal took up three entire TV channels, not just one. Secondly, a regular black-and-white TV set would be able to pick up either the red signal or the green signal or the blue signal — but not a balanced B & W signal. The existing B & W TV sets out in the market would need a special adaptor, to be able to extract a B & W picture from a colour signal.

There was also a potential maintenance problem. — a spinning Nipkow scanning disc, in both the camera and the TV receiver.

The official premiere of CBS colour broadcasting was on 25 June 1951. The one-hour special broadcast featured Ed Sullivan and other CBS stars.

The quality of the CBS colour system was superb. But while there were 12 million TV sets in the USA, only a few dozen of them could pick up the colour transmission. CBS was broadcasting, but nobody was watching. If nobody bought the colour TV sets on which to see the newfangled colour the new system was doomed.

Luckily for CBS, the RCA colour system (also using a three-colour system) looked terrible.

RCA Electronic Colour System

Back in September, 1950, there had been a test between the CBS and RCA systems. The headline in *Variety* read 'RCA Lays Colored Egg'. David Sarnhoff, the Chairman of RCA, said, 'the monkeys were green, the bananas were blue, and everybody had a good laugh'.

Soon afterwards, on 10 October 1950, the FCC approved the CBS colour system. Sarnhoff was in trouble. If the RCA system was not accepted by the NTSC, his company would lose a lot of money. Sarnhoff needed a colour TV system that was better than the current RCA colour system and the CBS colour system.

It would be entirely electronic, with no moving parts. It would be compatible with the existing B & W TV sets. It would also use up only as much bandwidth as a single B & W channel. The only problem was that such a system did not yet exist.

Sarnhoff threw himself, and the research labs of RCA, into a very expensive battle. Eight-hour work days expanded to 18 hours. He personally inspired the engineers and the scientists in the RCA labs to make a better colour TV system. He even authorised special $10 000 bonus payments for engineering breakthroughs.

The Clever New System

In a mighty six-month-long burst of inspired research in late 1950, the RCA people came up with the analogue TV system that we still use today. The three separate colour images, which previously took up three whole TV broadcast channels, were electronically squashed into one B & W channel. All the information was blended into two separate signals — luminance and chrominance.

Luminance = > Brightness and Sharpness

Luminance is the B & W signal. It gives you the brightness, and sharpness, of any part of the screen. The luminance signal is usually made up of 30 per cent red signal, 60 per cent green signal, and 10 per cent blue signal. When the luminance signal is picked up by a B & W TV, it appears as a nice clear image, with many subtle shades of grey. If you adjust all of the colour out of your colour set, you are looking purely at the luminance signal — the B & W signal.

Chrominance = > Colour

The chrominance signal carries two types of colour information. Phase modulation controls the hue of the colour — whether it's red or violet, or any colour in between. Amplitude modulation controls the saturation of the colour — whether it's a pastel or a deep vivid colour. (A pastel colour is just a vivid colour diluted with lots of white).

Both the chrominance and luminance signals can fit into one TV channel, because the peaks of the chrominance fit into the valleys of the luminance, and vice versa.

Colour Stalemate – RCA Wins

At the end of June, 1951 (just a few days after the unwitnessed premiere of the CBS colour system), RCA unveiled their all-electronic (and B & W compatible) system.

After a few months, both CBS and RCA were in a bind. CBS had the FCC's blessing, but nobody would buy their B & W-incompatible colour TV sets. It seemed silly to have two TV sets — one for colour, the other for B & W.

RCA had a B & W-compatible colour TV set, but it was not fully worked out — so they really didn't want to release it just yet. Furthermore, RCA was already making a fortune from selling B & W sets, and they didn't want to introduce colour sets until they had first saturated the market with B&W sets.

Conveniently for CBS and RCA, the Korean War intervened. The US government ordered that colour TV broadcasts, and colour TV set manufacture, be temporarily suspended from October 1951. Soon after, CBS pulled out of colour TV. RCA made another six million B & W TV sets, making it that much harder for CBS colour to be accepted. The RCA engineers used this time to improve their colour system.

NTSC Takes RCA System

On 17 December 1953, the FCC voted that the official colour system for commercial broadcast in the USA would be the RCA system.

By 1959, RCA had spent over $130 million developing and marketing colour TV — and had not yet made a profit. The turn-around began in 1960. David Sarnhoff took the *Walt Disney Show* from ABC to NBC, where it would be broadcast for the first time in full colour. Gradually, sales of colour TV sets began to pick up. By 1965, RCA had made a profit of $100 million — mostly because of colour TV.

Cardiopulmonary Resuscitation

The medical phrase for a 'heart attack' is a cardiopulmonary arrest. 'Cardio' means 'heart', 'pulmonary' refers to the 'lungs', and 'arrest' means a 'stoppage'. So a cardiopulmonary arrest means that the heart no longer pumps blood through the body, and the lungs no longer shove air in and out.

One thing that you can do for somebody who has had a heart attack is CPR. CPR as we use it today first appeared in the 1960s.

Polio Gave Us CPR

CPR came to us via polio. Today polio is a relatively rare disease, thanks to vaccinations. But in 1952 in Copenhagen, one single polio epidemic affected 5000 people. Five hundred of them could not breathe by themselves, because the polio had damaged the nerves involved in breathing. Ventilation tubes were pushed through surgical cuts in their necks, and they were ventilated by rubber bags that were squeezed by hand. All medical research and teaching was cancelled at the major Copenhagen hospitals. Medical students manually ventilated the patients 24-hours-per-day, 7-days-per-week. Eventually the crisis resolved, and the surviving polio patients could now breathe for themselves. Only 40 per cent (not the expected 85 per cent) of the 500 polio patients with respiratory paralysis died.

This single epidemic stimulated doctors to think about pulmonary resuscitation. The 'iron lung' was developed. It fitted the victim tightly at the neck and legs, and expanded their lungs by negative-pressure whole-body ventilation. The iron lung expanded the lungs, but what about the heart. How could you take over the job of the heart, once it had stopped?

If you see a medical drama film made before 1960, you might see a doctor resuscitating somebody by cutting open their chest, grabbing the victim's heart with their gloved hands and gently squeezing it. Of course, the long-term survival rates were virtually zero.

This very dramatic technique soon evolved into the CPR that we still use today. In 1960, Dr W.B. Kouwenhoven and his colleagues published their famous paper in *JAMA (Journal of the American Medical Association)*, describing how to do closed chest massage, or CPR.

CPR became very popular. By 1977, 15 million people in the USA had already learnt this procedure. It is most often used to keep somebody alive after their heart has stopped.

Why the Heart Stops

Each year in the USA, 800 000 people die from heart attacks. Many of these people die before they reach medical care and some of these deaths could have been prevented by CPR.

The most usual reason that the heart stops beating in an adult is that the normally regular rhythm has degenerated into ventricular fibrillation (VF) or ventricular tachycardia (VT). In fact, a recent study shows that sometimes a simple blow to the chest (such as from a ball) can set off VF, which will then stop your heart.

Your survival chances with CPR are highest when you have VT, lower when you have VF, and lowest of all when your heart has stopped entirely.

VENTRICULAR FIBRILLATION (VF)

The human heart has four chambers. The first two chambers (the right atrium and the right ventricle) pressurise the bluish low-oxygen blood (first to 10 mm of Hg, and then 25 mm of Hg) and then pump it into the lungs to be oxygenated. The blood pressure drops to around 10 mm of Hg in the lungs. The reddish oxygen-enriched blood flows from the lungs into the left atrium (15 mm of Hg), and then into the left ventricle (120 mm of Hg), and finally out into the general circulation.

The many individual strands of muscle that surround the two atria and the two ventricles have to fire in a very specific and co-ordinated order, so that they can efficiently push the blood from the heart into the general circulation. Usually, the muscles are so well co-ordinated that on each beat, about 80 per cent of the blood in the ventricles gets pushed out. These muscles are fed by the coronary arteries.

Ventricular fibrillation (VF) occurs when the muscles of the heart fire randomly. Instead of a smooth co-ordinated beat about once per second, it feels like a bag of worms all wriggling at random. The oxygen-rich blood sits in the lungs, and does not travel to the organs where it's needed. Once the brain has been starved of oxygen for 4–6 minutes, it dies.

Sudden heart stoppage is usually caused by VF. The causes of VF depend on whether the patient is inside, or outside, a hospital. When VF happens outside a hospital, it usually reflects a long-term blockage of coronary arteries (with fat) that has starved the heart muscles of oxygen, which eventually causes electrical instability.

Ventricular fibrillation (VF) can often begin as ventricular tachycardia (VT). 'Tachy' means 'fast', so 'tachycardia' means a fast heartbeat. The abnormal heartbeat can begin with a few short runs of VT, which get longer, and then degenerate into VF.

Even today, we do not understand what *causes* ventricular fibrillation. Even more surprising, we do not understand how the shock from the defibrillator paddles stops the unco-ordinated muscle twitching of fibrillation, and converts it back to a regular once-per-second beat.

CPR: P = PLUMBER'S PLUNGER?

In 1990, Keith G. Lurie and his colleagues wrote a paper to *JAMA* called 'CPR: the P stands for Plumber's Helper'.

It describes the successful resuscitation of a 65-year-old man with a plumber's helper — the stick with a rubber cup on the end, that can both push water down your drain, as well as suck it back. The 65-year-old man, who had severe coronary artery disease, collapsed in front of his family. His son was not well trained in CPR. He tried unsuccessfully to resuscitate his father.

Dr Lurie then writes, 'The son then remembered that his mother had resuscitated her husband six months earlier with a toilet plunger. Thus, the son ran upstairs, took out the plunger, and proceeded to plunge his father's chest for about 10 minutes until the paramedics arrived.' Dr Lurie then goes on to speculate that: '. . . the plunger served as . . . [an] effective chest compressor, but the suction between the chest wall and the plunger generated significant negative pressure and served to ventilate the patient as well. The son . . . recommended that we place toilet plungers next to all the beds in our coronary care unit. We recommended that he take a basic CPR course, but had to admit that it's hard to argue with success.'

But how can you ethically test such a device? It's hard to get informed consent from someone who is unconscious, or in great pain from their heart attack. Whatever the reason, the follow-up studies did not show that this plunger device was an improvement.

I guess we'll have to wait until further results come out, before we can all take the plunge.

DROWNING

The sooner you deliver CPR to a victim, the better their survival rate.

One study looked at 83 adults (average age of 31 years) who were admitted to a hospital for near-drowning, or drowning. All of those whose lungs only had stopped working, and 33 per cent of those whose lungs *and* hearts had both stopped, survived CPR to be successfully discharged. Factors that predicted survival included being young, being in the water for less than 10 minutes, not having swallowed any water, and a core body temperature less than 35°C (95°F), at the hospital admission.

Another study of 135 children and young adults under 20 years old found that the shorter the time underwater, the better the chance of survival. If they were submerged for less than five minutes, their death rate was 10 per cent. The death rate was 56 per cent if they were underwater for between six and nine minutes, 88 per cent if they were underwater for 10 to 25 minutes, and 100 per cent if they were underwater for longer than 25 minutes.

Dramatic CPR — ER, Chicago Hope, Rescue 911

CPR looks very dramatic. Quite a few TV medical shows use a bit of CPR to move the story along when things are getting a bit quiet.

A recent study in the *New England Journal of Medicine* had Susan J. Diem (from the Veterans' Affairs Medical Center in Durham, North Carolina) and her colleagues watching a lot of medical drama TV. They were looking for every single instance of CPR, the age of the patient, why they needed CPR, and what happened to them after the CPR.

They watched every single episode of the shows *ER* and *Chicago Hope* during the 1994/95 viewing season — 47 episodes in all. *ER* (Emergency Room) shows a busy emergency centre, while *Chicago Hope* deals with the very busy lives of a group of surgeons. The team also watched 50 consecutive episodes of the show called *Rescue 911* over a three-month period in 1995. *Rescue 911* shows re-enactments of medical incidents.

Altogether, there were 60 cases of CPR shown in these 97 episodes (11 on *Chicago Hope*, 18 on *Rescue 911*, and 31 on *ER*).

But there are a few problems with how these shows portray CPR.

It Isn't True

First, these three shows give a wrong impression of who gets CPR. In TV-land, young adults, teenagers and children received 65 per cent of the CPR handed out, but in real life, cardiac arrest/CPR happens much more frequently to elderly people.

Secondly, in TV-land, cardiac arrest usually happened because of some kind of injury that has nothing to do with heart disease, such as gunshot wounds or motor vehicle accidents. In TV-land, only 28 per cent of CPR was given to people who had some kind of heart disease. But in real life, about 85 per cent of cardiac arrests happen because the patient already had a previous cardiac condition.

Thirdly (and this is where the misinformation can be really dangerous), CPR in TV-land had an enormously high success rate. 75 per cent of patients who got CPR were alive an hour or so after it was given, while 67 per cent survived long term. In fact, on *Rescue 911*, the success rate was 100 per cent.

The average TV viewer would think that their chances of survival after CPR would be quite good. But TV is telling them lies.

Survival After CPR – 75% or 5%?

In real life, survival after CPR is not common. Of course, the survival rate varies enormously with the age of the patient, and the cause of the heart arrest. For example, you can see that there will be a big difference in survival rates between a 16-year-old boy whose heart has stopped because of a blow with a baseball, compared with an 85-year-old woman with breast cancer that has spread throughout her body, including her lungs, having a cardiac arrest.

In general, the long-term survival rate is between 2 and 30 per cent for arrest/CPR

'WOULD YOUR WIFE MIND IF SHE CAUGHT YOU KISSING ANOTHER MAN?'

This was the huge sign on the roadside billboards, set up as a campaign by St John Ambulance Service in New South Wales in Australia. This rather provocative advertisement was done deliberately to get people thinking about CPR and First Aid.

The small print at the bottom of the board said '*not if you've taken a First Aid course with St John*'.

inside a hospital, and between 6 and 15 per cent for arrest/CPR outside a hospital.

For your average *elderly* patient, the chances of *long-term* survival after a heart attack (which happens *outside* a hospital), is less than 5 per cent. The reported survival rates after trauma/cardiac arrest/CPR vary from 0 to 30 per cent.

Life, Death, and the Other . . .

Finally, the possible results of CPR are not just the two alternatives of a complete and full recovery, *or* death. There are some very distressing in-between possibilities. You could be left with prolonged suffering, severe brain damage, or a protracted and undignified death.

CPR

Everybody should know how to do CPR. The *cardio* bit means pressing on the chest at the right place, and with the right timing. If you do this correctly, you can push blood out of a non-beating heart around the body. The *pulmonary* bit refers to the mouth-to-mouth breathing to get fresh air into the lungs. *Resuscitation* means to bring back to life.

CPR is an emergency procedure. Its purpose is to restore normal breathing and circulation after an emergency such as a drowning or a heart attack. CPR should be performed only on people who are unconscious, and who are not breathing and/or have no heartbeat.

FOLLOW THE ABC OF CPR:

First, open and then clear the airways of the victim. Do this by placing the victim on their back on a flat rigid surface, remove any foreign matter from their airways (eg. TV dinners or vomit) and tilt the head back so that the chin is pointing to the sky. This is the Airways or 'A' part of ABC.

The 'B' part of ABC, is Breathing. Clamp the victim's nostrils shut with your fingers, make an airtight seal between your mouth and theirs (either by direct mouth-to-mouth contact, or by a bridging piece) and breathe forcefully into their lungs around 12 times per minute. The air that comes into your lungs has about 20 per cent oxygen. The air coming out has about 16 per cent oxygen, which is enough to keep the unconscious person alive.

The third step is the 'C' of ABC — *Circulation*. Check for a pulse, usually on the carotid artery in the side of the neck. If there is no pulse, compress the lower part of the rib cage, in the middle, by about 4–5 cm (1.5–2 inches). Do this 80 times per minute.

The normal heart will pump about five litres of blood per minute. During a standard manual CPR, it will pump between 25 and 50 per cent of normal blood flow. It's not much, but it's enough to keep the person alive.

There is the possibility of a Chronic Vegetative State, where the patients cannot walk, talk or feed themselves. Some studies had found that the number of patients in a chronic vegetative state after CPR is about 2 per cent. This state usually lasts until the patient dies. Some people believe that a Chronic Vegetative State is worse than death.

These other options were shown only once in all the episodes of the medical TV shows surveyed. There was the case of a 16-year-old boy who had inhaled butane and a cleaning agent, had a cardiac arrest, but was resuscitated with CPR. In that single case, the boy was shown as having only a mild speaking disability. However, he still had a normal walk, had completed high school, and had become a successful motivational speaker who spoke out against the evils of drug abuse!

Must We Always Use CPR?

A paper by Leslie J. Blackhall is called 'Must We Always Use CPR?'. He says that CPR is simply not appropriate for some diseases, and some patients. He considered the case of a young woman who had leukaemia that had not responded to chemotherapy. Her bones were riddled with further cancerous growths, and her lungs were affected by a virulent and rapidly progressive pneumonia.

He writes that, as far as the hospital staff was concerned, 'the choice was clear: death on the oncology ward, surrounded by family members and the nurses and doctors who knew the patient well, versus death in the intensive care unit after multiple, invasive, painful and dehumanising procedures.'

However, the young woman and her relatives had not *seen* it all happen dozens of times before. He writes, 'they were not well informed about the likely outcome of CPR. They had never been in an intensive care unit, or seen a respirator. For them, the choice (of CPR) appeared to be between a chance of life and certain death. When they chose CPR, they were actually choosing something that did not exist — a chance for the patient to live.'

However, in one study, if the hospital staff did carefully explain the many options of CPR to aged patients, many of them specifically asked that they *not* get CPR, in the event of a cardiac arrest.

Faith and Miracles

But there was another subtle misrepresentation of truth in TV-land. In *Rescue 911*, the doctors would nearly always predict a very bad outcome for the patient. When the patient did (as they always did) survive, the doctors would then say that it was a 'miracle'.

It is very important to have faith. But if you believe that a miracle will always come along to fix you, sometimes that faith can actually harm you.

An 85-year-old woman in terrible pain with cancer-of-the-everything might not want CPR if she were to have a cardiac arrest. Perhaps she would prefer to die naturally and quickly, rather than to be revived with CPR, and *then* spend the remaining two weeks of her life with brain damage, with tubes into her body, and a slow and painful death. Patients need to know that medical procedures like CPR are useful in some circumstances but that they will not produce miracle recoveries.

After all, everyone eventually dies. Sometimes it is more painful to postpone the inevitable. It then becomes a matter of *how* we die.

Why Medical Truth Matters

If I were to quietly say the word 'fire' in a room around a dinner table, it wouldn't cause any great distress. But if I were in a movie theatre after the houselights had failed, and I shouted out the word 'fire' very loudly, and three people were trampled to death in the rush, then I would be responsible for those deaths.

In the same way, people who write about medical matters must be aware of their responsibilities. Providing wrong information about medical matters has a different significance and consequence from providing wrong information about gardening.

Medicine is moving into a new stage in which patients are becoming much more informed, and where patients and their doctors share the decision-making, and the responsibility for what happens. But one study showed that 92 per cent of patients who were older than 62 years got most of their information about CPR from television. It is difficult for them to make an informed decision about CPR when the information they get from TV is misleading or simply incorrect. Patients now expect doctors to be much more open and communicative about treatments, options, and side effects. Doctors can use this opportunity to provide accurate information, and to correct misinformation.

But there is one important lesson from TV-land. If you're going to have a cardiac arrest, make sure that you're on the set of one of these TV shows where miracles *do* happen.

Charles Anderson, 'Violence in Television Commercials During Nonviolent Programming — The 1996 Major League Baseball Playoffs', *Journal of the American Medical Association*, Vol. 278, 1 October 1997, pp 1045–1046.

Rob Ashton, 'Don't just lie there, turn on the telly', *New Scientist*, No. 2021, 16 March 1996, p 7.

Neal A. Baer, 'Cardiopulmonary Resuscitation on Television — Exaggerations and Accusations', *New England Journal of Medicine*, Vol. 334, No. 24, 13 June 1996.

Leslie J. Blackhall, 'Sounding Board — Must We Always Use CPR?', *New England Journal of Medicine*, Vol. 317, No. 20, 12 November 1987, pp 1281–1285.

Tony Charlton, 'Prevalence of Behaviour Problems Among 9–12-year-old Pupils on the Island of St Helena, South Atlantic: Preliminary Findings', *Psychological Reports*, Vol. 74, June 1994, p 890.

Tony Charlton, David Coles and Tessa Lovemore, 'Teachers' Ratings of Nursery Class Children's Behaviour Before and After Availability of Television by Satellite', *Psychological Reports*, Vol. 81, August 1997, pp 96–98.

Susan J. Diem et al, 'Cardiopulmonary Resuscitation on Television — Miracles and Misinformation', *New England Journal of Medicine*, Vol. 334, No. 24, 13 June 1996, pp 1578–1582.

David E. Fisher and Marshall Jon Fisher, 'The Color War', *Invention and Technology*, Winter 1997, pp 8–18.

Keith G. Lurie et al, 'Evaluation of Active Compression-Decompression CPR in Victims of Out-of-Hospital Cardiac Arrest', *Journal of the American Medical Association*, Vol. 271, No. 18, 11 May 1994, pp 1405–1411.

'Violent Commercials Common During Baseball Playoffs — Children Can See up to 14 Violent Commercials During One World Series Game', *Science News Update*, 1 October 1997.

Richard Wiseman, 'The megalab truth test', *Nature*, Vol. 373, 2 February 1995, pp vii, 391.

YOUR NAME IS YOUR JOB

MS RICA MORTIS
TAXIDERMIST

Some so-called primitive societies believe that your true name is a 'thing of power'. They believe that your true name is a very serious thing and should not be trifled with. So these people have a secret name (which is kept very secret), and a common name (which everybody else knows, and which is what they are generally called by).

But in today's sophisticated culture, how powerful is your surname? Can it influence your life?

NAMES - MOST, SHORTEST, LONGEST

According to *The Guinness Book of Records*, the person with the *greatest number of first names* is Laurence Watkins, who was born on the 9th of June, 1965, in Auckland in New Zealand. He has 2310 first names, which he legally added, by deed poll, in 1991.

The *shortest family name* is a single letter. Every letter of the alphabet has been used as a family name, except for 'Q'. Worldwide, the most popular 'letter-name' is 'O', which is especially common in Korea.

The *longest first name* is 1019 letters long. It belongs to the daughter of Mr and Mrs James Williams. She was born in Beaumont in Texas, on 12th September, 1984. On her birth certificate, she was called 'Rhoshandiatellyneshiaunneveshenk Koyaanfsquatsiuty Williams', but shortly after her birth, on the 5th October, her father filed paperwork that gave her longer first and middle names.

Names Carry Power

For a long time it has been thought that names carry power.

In the Bible, the Book of Genesis 2:19 tells how God guided Adam in naming all the animals, and presumably, getting power over them: *Now the LORD God had formed out of the ground all the beasts of the field and all the birds of the air. He brought them to the man to see what he would name them; and whatever the man called each living creature, that was its name.*

A similar story is given in *Cratylus*, one of the dialogues of the Greek philosopher, Plato. Socrates tells how the gods were the first to give names to things. Socrates also tries to decide if words have meanings just because we humans decide it is so, or if they intrinsically have meanings.

In some societies, knowing the name of some object, animal or person gives you power over it or them. Even today, in some societies there are taboos against mentioning the name of a recently departed relative.

The North American Indian tribe called the Navajo have a special 'fourth person' in their grammar. This lets them talk to, and mention, another person who is within hearing range, without actually naming that person. This special grammar helps them avoid invoking the power that a 'name' has over another person.

Do Names Reflect Your Job?

Back in October 1994, Jen Hunt wrote an article in *The Psychologist* called 'The Psychology of Reference Hunting'. She basically gave sensible advice on how to keep up to date with the reading that you had to do. She even dealt with how to photocopy successfully — advising that you should avoid exam time when the photocopiers are super-busy, and wear sunblock and sunglasses to protect you from the bright photocopier light.

YOUR NAME IS NOT YOUR JOB?

According to the *Encyclopaedia Britannica*, last names came from people's jobs, locations, achievements, personal appearance, their father's name, or 'hypocoristic' names. 'Hypocoristic' names are just personal, or intimate or shortened names, such as Tom (from Thomas), Jim (from James) or Hal (from Harry). The *Encyclopaedia Britannica* lists many of these names.

Names that come from jobs or professions include Mason, Butcher, Baker, Hooper, Weaver, Clerk, Clark, Clarkson (the son of a clerk), and Bundy, Bound, Bonds, Bond (bondman). Jewish family names that have a religious origin include Löwy, Levi, Halévy, (tribe of priests), Kantorowicz, Cantor, Canterini (under priest), and Kaan, Kahane, Cohen, Cahen, Kohn (priest).

Names that came from appearance include the Russian name, Krasnoshtanov, which means 'red pants'. After the Revolution, many such subservient names were changed to more noble names, like Orlov, which means 'eagle'. The French names that come from 'red hair' include Leroux, Roujon, Roux, Rousseau, Lerouge and Roussel.

Names that came from a father's name include Smithson, Johnson, Clarkson (also related to a job), Richardson, Dickson (Dixon, and Dickinson), Harrison, Henderson, Gibson and Gilbertson.

In England, 'Fitz' means 'son of' — which originally came from the Norman French word 'fis', which means 'son'. This gives us names such as Fitzgerald and Fitzsimmons.

In Scotland, 'Mac' or 'Mc' means 'son of' as well. So McDonald means 'son of Donald'.

In Ireland, the prefix 'O' also means 'son of', which gives us names such as O'Brien. And in Wales, 'P' means 'son of', which gives us Powell, which means 'son of Howel'. In Greece, 'poulos' is added to 'Dimitrios' to give 'Dimitriopoulos', which means 'son of Dimitrios'.

A 'patronymic' is what you call a name that comes from a father's name. In Russia, everybody has a patronymic, as well as their first name and last name. This patronymic depends on the sex of the child, so my son would be called Karlovich, while my daughter would be called Karlovna.

'Hypocoristic' names are personal names and they include nicknames, such as Little, Biggs, and Grant (large or grand). Many family names come from these intimate names. So 'Gilbert' led to Gibbons, Gibbs, Gipps, Gilpin, and so on, while 'Gregory' led to Gregg, Grigg, Greig, and so on.

But the origins of some names have been lost. For example, where did these Czech names come from? Skocdopole, which means '*Jump into the field*', and Nejezchleba, which means '*Don't eat bread*'?!

She covered many ideas in her article, but the one that struck the fancy of the scientific world was 'Nominative Determinism' which discussed how a person's name sometimes reflects their job. Jen Hunt claimed that, in some cases, '*Authors gravitate to the field of research which fits their surname*'.

Soon, the *New Scientist* magazine was swamped with examples of people whose names suited their jobs. This is a little list that I have gathered from the *New Scientist*, and other sources, over the last few years. I am sure that it would be very easy to compile a much bigger list.

NAMES & JOBS

ORGANISATIONS	LOCATION/PEOPLE
British Meteorological Office	Have staff named Flood, Frost, Thundercliffe and Weatherall
Clearwater Filter Systems	Ms Tapp
Fine Cut Ltd	In Lancing, Surrey
Gloster Aircraft — Aeroelasticity Department	Cecil Partridge, Terry Heron, Pat Woodcock & Harry Peacock
Kennel Club	In Barking
London Hair Restoration Clinic	Situated in Wigmore Street
Royal Society of Chemistry's Water Chemistry Group	Hugh Fish
Royal Society for the Protection of Birds	Mark Avery
US National Weather Service	Dave Storm
Urban Bird Control	Employs John Sparrow, Graham Crowe, Grant Parrott and Nigel Hawkes

INVENTORS	INVENTION
Karen Bond	Adhesive compositions
Monsieur L. Cock	A new kind of condom
Mr Skidmore	A tubeless tyre
Paul Coffman	A tobacco filter
Stuart Filhol	A means of dental anchoring

WHAT WILL WE CALL BABY?

Most parents have a bit of trouble in choosing a name for their newborn child. That's why books of names are so popular. But God made it easy for Mary and Joseph, the parents of Jesus.

In the Bible, Luke 1:26 to 1:31 says: 'In the sixth month, God sent the angel Gabriel to Nazareth, a town in Galilee, to a virgin pledged to be married to a man named Joseph, a descendant of David. The virgin's name was Mary. The angel went to her and said, "*Greetings, you who are highly favoured! The Lord is with you.*" Mary was greatly troubled at his words and wondered what kind of greeting this might be. But the angel said to her, "*Do not be afraid, Mary, you have found favour with God. You will be with child and give birth to a son, and you are to give him the name Jesus.*"'

Alan Pee	Chesspool District Council's cesspool emptying service
Ashley Burns	Fireman
Cardinal Sin	Archbishop of Manila
Con Allday	Public relations officer at Sellafield
David Butcher	Executive Director of the Royal Society for the Prevention of Cruelty to Animals in NSW
Douglas Dick	Product manager for a pharmaceutical company producing a treatment for penile warts
Dr Moneypenny	Lecturer in banking and finance
Francis Fry	Works on the effects of radiation exposure
Gene Shearer	Biologist — US National Institute of Health
Geoffrey Gold	Editor in chief, *Australian Journal of Mining*
Ian Forest	Works for the Forestry Authority in Midlothian
Jim Shonk	Real Estate Agent
John Death	Civil Aviation Authority's public relations person in Sydney
John Dickie	Australia's Chief Censor
John Fish	Marine Biologist at Aberystwyth University

Judge Hanger	Queensland District Court Judge
Katherine Hacker	Manager of anti-virus computer software
Katherine Cable	University of Sheffield telecommunications officer
Liz Peace	Defence Research Agency's spokeswoman
Mark Quickfall and Tanya Pickup	Scenic joy flight operators in Auckland, New Zealand
Mr Lust	Sex therapist
Mr Vice	Police officer in anti-porn office
Neil Gamble	Chief Executive of the Sydney Harbour Casino
Professor Michael Lean	Professor of Nutrition at the University of Glasgow, who spoke at a conference on obesity
S. & M. Grocock	Manufacturers of artificial limbs
Sue Pipe	General secretary of the Industrial Water Society
Warren Breeding	Manager of the South Australian Station which was overrun by rabbits
Will Drown	Sales manager of Crewsaver Company who make life jackets

AUTHOR/S	BOOK TITLE
A. Quick, V. Browne & S. Fox	Co-authored a paper in the journal *Surface Science*

(You might remember the old typing exercise that uses all the letters of the alphabet in just one sentence, 'The quick brown fox jumps over the lazy dog'.)

AUTHOR/S	BOOK TITLE
A. J. & H. A. Barker	*The Complete Book of Dogs*
A.M. Glass	*Optical Materials*
David Steele	*The Chemistry of Metallic Elements*
David Killingray	*The Atom Bomb*
Derek Heater	*The Cold War*
Doug Stone	*How to find Australian Gemstones*
Edgar E. Mountain	*Geology of Southern Africa*
Estelle Fuchs	*Life, Love & Sex*
F. H. Rainwater & L. L. Thatcher	*Methods for Collection and Analysis of Water Samples*
Freda Dinn & Paul Sharp	*The Observer's Book of Music*
G. M. Flood	*Sewage Disposal from Isolated Buildings*
George Gamow, Ralph Alpher and Hans Bethe	Wrote a paper on the Big Bang and cosmic background radiation

(Gamow and Alpher did most of the work, but they went out of their way to make their colleague, Hans Bethe, a co-author. This meant that the author list became, in alphabetical order, Alpha, Beta and Gamma — which are the first three letters of the Greek alphabet!)

AUTHOR/S	BOOK TITLE
George Plumptree	*Great Gardens, Great Designers*
Gladys Elder	*The Alienated: Growing Older Today*
J. Angst	*Bipolar Manic-Depressive Psychosis*
James Reason	*Absent-Minded?*
M. Bedrock	Wrote a thesis called *Sedimentology of some Westphalian C sequences*
Raymond Bush	*Tree Fruit Growing*
Richard Trench	*London Under London — Subterranean Guide*
Sophie Wormser	*About Silkworms and Silk*
Stephen Fisher	Editor of *Man and the Maritime Environment*
T. L. Vincent, M. V. Van & B. S. Goh (Vincent van Gogh!)	*Ecological stability, evolutionary stability and the ESS max. power principle*

(In the above article, Vincent (from the University of Arizona) really did work with Goh (from the University of Western Australia). Van was a 'visitor' to Vincent's department in Arizona, and actually didn't do much of the work at all, but, as Vincent admitted in the *New Scientist*, 'We couldn't pass up the opportunity'.)

AUTHOR/S	BOOK TITLE
W. Frost	*Heat Transfer at Low Temperatures*
Walter Russell Brain	Former editor of *Brain*

Able, Best & Brilliant	Fortunate doctors' names	Dr Misri	Psychiatrist specialising in depression
C. L. Chew	Expert in the musculature of the jaw	Dr Rash	Dermatologist
		Dr Surgeon	Lives in America
Chris Pullin	Dentist from Sydney	Drs Weedon & Splatt	Urologists
Croak, Perish & Klutz	Unfortunate doctors' names	F. A. Payne	Dentist from New Zealand
Dr Bone	Orthopaedic Surgeon		
Dr Blewett	Not a medical doctor, but was once the Minister for Health in Australia	Geneticist E. Tatum	This geneticist's name spelt backward is 'mutate'
		John Roger, Jim Cummins, Bill Breed & Karen Mate	All attended a spermatology conference
Dr Couch	Psychiatrist		
Dr Leanne Craze	Wrote Commonwealth *Peak Mental Health Body Report*		
		Mr & Mrs Screech	Dentists, their practice is called Dentith & Dentith
		P. A. Gummers	Dentist from Glasgow
Dr Cutter	Delivered Dr Craze's baby by caesarean section	Roy Phang	Dentist
Dr Doctor	There are 18 in America	Sister Blood	Male nurse from Sydney
Dr Ernest Reginald Crisp	Senior radiologist from Melbourne	Y. T. Chew	Dentist from Ashford Kent
Dr Gass	Anaethetist		
Dr Grunt	Animal behaviourist		
Dr Hart	Cardiologist		
Dr J. Pupil	Optician in a small village in South of France		
Dr Graeme Killer	Medical officer to Australian ex-Prime Minister Keating		
Dr Knee	Rheumatologist		

DR HUGH GARSE
LIPOSUCTION SURGEON

NOMINATIVELY DETERMINED DEATHS

Dennis Christian	Called on God to uphold him as he jumped from his 13th-storey balcony
Joachim Feller	Sleepwalked through his bedroom window
Stephen Duck	An 18th-century poet who drowned himself

RANDOM SIKH NAMES

Among some Indian Sikhs, the baby is brought to a Sikh religious person when it is a few days old. A religious book, the *Adi Granth*, is opened at random. The name given to the child begins with the first letter of the first word that is found on the left-hand page.

NAMES OF POPES

When a man ascends to being the Pope of the Roman Catholic Church, he usually abandons his given and family names. He then often takes on the name of someone who embodies his ideals. So the name of Pope Paul VI, came from St Paul, who travelled much in his missionary activities in the first century A.D.

REFERENCES

Bob Bagnall, 'Alas Smith, Jones and Lewis', *New Scientist*, No. 1795, 16 November 1991, p 46.

Encyclopaedia Britannica, 1996 edition.

'Face Facts: The Game of the Name', *Sydney Morning Herald*, 2 December 1994, p 22.

Bryan Gaensler, 'Vincent's Van', *New Scientist*, No. 2063, 4 January 1997, p 45.

Fiona Harari, 'When Your Name Can Drop You in It', *Australian*, 2 June 1993, p 17.

Jen Hunt, 'The Psychology of Reference Hunting', *The Psychologist*, October 1994, p 480.

New Scientist, No. 1956, 17 December 1994, p 64.

New Scientist, No. 2058, 30 November 1996, p 64.

New Scientist, No. 2080, 3 May 1997, p 64.

A.J. Splatt and D. Weedon, 'The Urethral Syndrome: Experience with the Richardson Urethroplasty', *British Journal of Urology*, Vol. 49, 1977, pp 173–176.

COLD BATHS, OLYMPIC GAMES AND HOT BODIES

Cold baths have never had a good name. People are usually given cold baths as a punishment, to cool their lustful impulses, or to sober them up.

People rarely take a cold bath voluntarily, but cold baths can be good for you. They can help you get through winter without feeling the cold, and they can speed up long-distance runners. They do this by changing your body temperature (for a while).

The Temperature Of Humans

What temperature is a human being supposed to be? The standard 'answer' is that the 'normal' temperature of a human being, when measured with an oral (mouth) thermometer, is 37°C (98.6°F). This answer is based on the observations made by a certain Dr Wunderlich over 120 years ago.

Until recently, everybody believed Dr Wunderlich. But (like practically everything in the universe) it's more complicated than that.

Your normal temperature is *not* a single temperature (like 37°C or 98.6°F). It actually varies during the day. Today, we say that the normal temperature range for a human, as measured with a medical thermometer in the mouth, is 36.4° to 37.2°C (97.5° to 98.9°F).

It's lowest in the morning at 6 am, and highest in the afternoon at 4 pm. However, your mouth temperature is actually 0.6°C (1°F) cooler than your central core temperature.

The temperature measured with a rectal thermometer is generally taken to be as good a reading of the core temperature as you will easily get.

A woman can measure her temperature each morning to work out when she is ovulating. The average morning temperature (as measured at 6 am) usually sits at a certain value for the two weeks before ovulation happens. But at ovulation, the temperature usually jumps by around 0.6°C (1°F). The temperature remains at this elevated level until menstruation occurs.

Hypothalamus = Thermostat

How does the hot water in your home plumbing system maintain its temperature? It is because hot water heaters have a thermostat. A thermostat has two parts — a device to measure temperature, and a switch. When the temperature gets too high, the thermostat switches the heating system off, stopping the water from getting too hot. When the temperature drops far enough, the thermostat switches the heating system back on.

Similarly, your body has a thermostat to keep its temperature constant. This thermostat is in the hypothalamus, deep in your brain. 'Hypo' means 'under', so the hypothalamus is under your thalamus (another part of your brain).

There are at least two separate methods by which your hypothalamus monitors

HYPOTHALAMUS = FATOSTAT

The hypothalamus does many things. Besides co-ordinating the release of many different hormones into the bloodstream, and being a thermostat, it's also a 'fatostat'.

When you lose weight, the hypothalamus increases your appetite and slows your metabolism, so that you store more fat. The hypothalamus tries to keep you at the previous level of 'fatness'.

SAGITTAL SECTION THROUGH BRAIN.

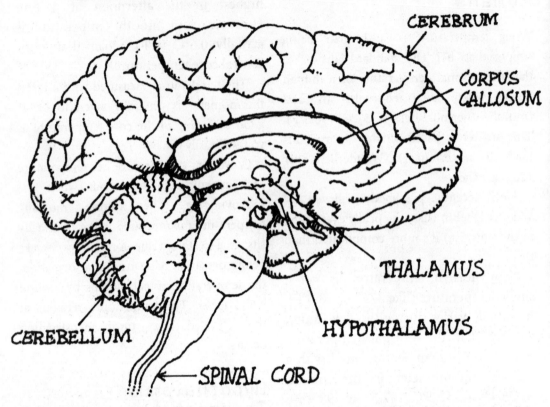

CEREBRUM

CORPUS CALLOSUM

THALAMUS

HYPOTHALAMUS

CEREBELLUM

SPINAL CORD

your 'temperature':

First, it receives signals from nerves in your skin. It averages all of these, and comes up with a certain temperature value for your body.

Second, the hypothalamus has local temperature sensors. It is able to sense the temperature of the blood that actually washes over itself, deep in your brain.

The hypothalamus processes these signals to come up with your current temperature. It then uses this information to set the balance between how much heat you make, and how much heat you get rid of.

DITCH YOUR DOONA

Dr Hugh Molloy (who, by a coincidence, was one of my teachers at the Children's Hospital in Sydney) claims that the Scandinavian doona is too hot for the warm Australian climate.

He says that the claims of doona manufacturers that they are 'warm in winter and cool in summer' are ridiculous. He claims that some doonas are equivalent to six blankets and he asserts that doonas will overheat your body, interfere with your deep sleep, encourage skin disease and even make your hair go lank.

How You Make Heat

When you rest in a chair, much of the heat that you generate comes from the normal metabolic activity in your liver and heart. But when you exercise, most of your heat comes from your skeletal muscles. In exercise, you can increase the amount of heat that you generate by a factor of 20. This can sometimes overwhelm your mechanisms for dumping heat, such as on a very hot and humid day.

When you're cold, you generate body heat by increased cell metabolism, shivering, and by moving your muscles (getting active). Your body conserves internal heat by closing down the blood vessels in your skin, and by triggering certain behaviour (such as prompting you to put on more clothes).

COLD WEATHER AND TESTICLES

Back in 1982, Dr Shukla published a paper called 'Association of cold weather with testicular torsion'.

'Testicular torsion' means that one of the testicles has twisted. This is a surgical emergency. If the testicle remains twisted, the blood supply to that testicle can stay closed off. As a result, the testicle will die after about six hours.

Testicular torsion is uncommon. It does not happen in a normal fully descended testicle — there has to be some minor abnormality of the testicle. It happens most commonly to males between the ages of 10 and 25, while the second most common age group is during infancy.

The most common symptom is that the patient suddenly has an agonising pain in the groin and lower abdomen, and then vomits. Sometimes the testicle can be untwisted by hand, if you can treat it in the first hour. But usually the testicle has to be untwisted surgically.

Dr Shukla's paper looked at 46 patients who, over a five-year period, had turned up at two Dublin hospitals with testicular torsion. In 40 of the 46 patients, the testicles had twisted when the temperature was less than 2°C.

It's fairly obvious why testicles should twist more at low temperatures. First, there is contractile tissue in the capsule of the testicle which contracts when the temperature is low, and can sometimes make the testicles align slightly wrongly. Second, the cremaster muscle in the scrotum will also contract in the cold and further contribute to the problem.

There was a young man from Dublin,
Whose testes were always a'troublin,
When the temperature did drop,
His testes did (k)not,
And he cried until they were undone again.

CHEMICALS CAUSE FEVER

A fever is a temperature higher than normal. It shows that some kind of disease process is happening in the body.

The boundary between a 'normal' and a 'high' temperature is a bit fuzzy. The maximum normal oral temperature will depend on the time of day. For 99 per cent of average adults, the upper limit of normal is 37.2°C (98.9°F) at 6 am, and 37.7°C (99.9°F) at 4 pm. A fever of 40°C (104°F) means that something quite serious is happening in your body.

In a fever, the thermostat in your hypothalamus is reset to a temperature higher than normal. Your fever is caused by chemicals. These chemicals are called pyrogens. '*Pyr*' means '*fire*', and '*gen*' means '*to make*'.

There are two kinds of pyrogens — exogenous (from outside your body) or endogenous (produced by your body). It was only as recently as 1948, that Dr Paul Beeson showed that you had a fever because your endogenous pyrogens had been brought into action (sometimes after being triggered by exogenous pyrogens).

Most exogenous pyrogens are bacteria, or the chemicals that bacteria make (such as toxins, or part of their bacterial bodies). For example, gram-negative bacteria make a pyrogen called LPS (Lipopolysaccharide). Gram-positive bacteria make pyrogens such as Lipoteichoic acid and Peptidoglycans. Very tiny amounts of these chemicals in your bloodstream will give you a fever.

Endogenous pyrogens are amino acids (a bunch of amino acids make a protein) made by various cells in your immune system. These chemicals then enter the blood circulation. Typical endogenous pyrogens include IL–1a, IL–1b, interferon A, and tumour necrosis factor (TNF). IL–1a and IL–1b are very powerful. A dose as little as five billionths of a gram per kilogram of body weight will give you fevers up around 39°C. The pyrogens IL–1a and TNF appear in many species, and probably evolved some 300 million years ago.

How You Dump Heat

Your heat is dumped into the environment at the surface of your body. About 90 per cent goes out through your skin, while the lungs get rid of the remaining 10 per cent. At rest, you dump about 70 per cent of your heat by *radiation*, and about 30 per cent by *evaporation* (that is, 'insensible' perspiration, which you usually don't notice). Not much heat vanishes through *convection* or *conduction*.

Water conducts heat away from your body about 25 times better than air does, so cold water will give you hypothermia much more rapidly than cold air at the same temperature. ('*Hypo*' means '*under*', and '*thermia*' means '*temperature*', so hypothermia means that your body temperature is too low.)

How You Regulate Heat

Normally, you produce more heat than you need to keep your body temperature around 36.8°C (98.2°F). So most of the time, your hypothalamus is working out ways to dump excess heat from your body.

The hypothalamus has a few options to regulate your temperature — basic and sophisticated. Suppose that you're too cold. On a basic level, the hypothalamus can act directly onto your muscles to make you shiver (the friction of the different muscle fibres rubbing over each other generates heat). This can increase your heat production from 50 kcal/hour to 300 kcal/hour.

The hypothalamus will also act on the more sophisticated part of your brain (the cerebral cortex) to make you do quite complicated behaviour — such as getting into shelter out of the cold weather, putting on warmer clothes, or (when you're in bed) curling yourself up into a ball.

Higher Level on Thermostat → Fever

We all know that a 'fever' is a higher-than-normal temperature, but something as 'simple' as measuring a body temperature can be quite tricky.

The temperature in the armpit may be 1°C (1.8°F) lower than your core temperature. This is because the blood vessels sometimes close down in a fever.

The oral temperature can be falsely low, because the moist air moving rapidly over the thermometer can cool it down.

When you have a fever, the hypothalamus is reset to a higher temperature, for example 40°C (104°F). You

'feel' cold, so you shiver to generate heat and huddle under the blankets to conserve heat. Once the blood bathing the hypothalamus and your skin has warmed enough, the hypothalamus will tell you to stop shivering. The hypothalamus will then maintain your temperature at that new higher level.

Chemicals such as aspirin or paracetamol (acetaminophen) will, via various chemical pathways, reset the thermostat in your hypothalamus down to a lower temperature.

Ferrets and Fish Favour Fever

Some scientists claim that 'Fever' has been around for hundreds of millions of years. Many different species have fevers — even fish and lizards. If you inject them with bacteria, the fish and lizards will raise their

WINTER HEART ATTACKS

A statistical study analysed 259 891 cases of sudden heart attacks (acute myocardial infarction). This study found that there were 53 per cent more heart attacks in the winter months than in the summer months. Surprisingly, it also found the same effect in both the warmer and colder states of the USA.

A paper in *The Lancet* claims that when you get a cool change coming through, people in warmer climates die more frequently than people in colder climates.

They measured the death rate for each 1°C fall in temperature below 18°C. They found that the death rate was greater in warm climates than in colder climates.

There were several possible mechanisms by which the cold could have killed people.

First, the cold could have made the blood more concentrated, which could have caused blockages in arteries — especially around the heart.

Secondly, the cold could have caused spasms in the coronary arteries, which could have ruptured little fatty (atheromatous) plaques. This would have caused a very rapid death due to heart attack.

Finally, the cold could have released stress hormones, which suppress your immune responses. With a lower immune response, people would be easily killed by a respiratory infection.

And why should the death rate be greater in a warm climate? Probably because people who regularly have a severe winter know how to dress for the winter, but people who usually have a mild winter simply don't dress properly.

temperature by swimming to warm water, or basking in the sun.

In animals, it seems as though fever is good for the creature. For example, it stimulates the immune system, and slows down the growth of several types of bacteria.

Ferrets that have been infected with influenza, will shed fewer influenza virus particles into their environment when they have a fever. Thus the fever seems to protect their fellow ferrets from infection.

Why Have Fever?

However, the benefits of fever in humans are not so well understood. There are very few medical scientists prepared to say that fever is actually good for humans (I think further research will probably show that a slight fever is good).

Some bacteria (*Gonococci* and *Treponema pallidum*) are sensitive to heat. In experiments using animals, artificial fevers have been generated that have killed these bacteria. But natural fevers in the human body never get to a temperature high enough to kill these particular bacteria.

Fevers Have Their Costs

When the temperature goes up by 1°C (1.8°F), the oxygen demand goes up by 13 per cent, and the heart rate rises by 8.5 beats per minute. A fever will also increase your need for fluids and calories.

In a fever, you break down your skeletal muscles. The big chemicals in the muscles turn into smaller chemicals. These breakdown products give you energy, as well as essential amino acids (which are then used, by your immune system, to make various chemicals to help you fight that fever).

About 3 to 4 per cent of children between the ages of one and five may have a convulsion if they have a high fever (around 39°C (102°F). However, this does not mean that the child is more likely to have epileptic fits in the future.

Fevers can be dangerous to unborn babies. If a pregnant mother has one single episode of a fever reaching temperatures over 37.8°C (100°F) in her first three months of pregnancy, the risk of the baby having a neural tube defect (such as spina bifida) doubles.

Cold Baths 'Reset' Thermostat for Winter

Now let's try to understand how cold baths can get you through winter.

The story begins in 1982, when two Canadian physiologists discovered that you can 'pre-condition' your body to tolerate the cold by taking a few cold baths. The result is that when winter comes, your body temperature actually drops, but you feel fine!

The physiologists were Drs M.W. Radomski (from the Institute of Environmental Medicine, Downsville, Ontario) and C. Boutelier (from the School of Physical and Health Education, University of Toronto). They did their study with 11 male volunteers who were scheduled to spend 16 days in the Arctic, living and sleeping in unheated tents — at an average outside temperature of −27°C (−17°F)!

The Cold Baths Pre-Conditioning

Only three of the 11 volunteers were brave enough to spend 25 to 40 minutes a day in a still bath at 15°C. They did this for nine days with their time in the cold water gradually increasing over the nine days. They left the water either when they felt too cold, or when their rectal temperature had dropped to 35°C (95°F).

The theory was that roughly half an hour a day in a cold bath might be enough to trick their bodies into pre-adapting to the Arctic freezer they were going into.

Nude Cold Tolerance Test

Fifteen days after the last cold bath, Radomski and Boutelier gave all eleven of the Arctic adventurers a standard Nude Cold Tolerance Test. In this test, you lie quite still, nude, in a room at 10°C (50°F) with calm air.

HYPOTHERMIA

Hypothermia implies that your core body temperature is lower than 35°C (95°F). Your hypothermia can be *mild* 35°C–32°C (95°F–89.6°F), *moderate* 32°C–28°C (89.6°F–82.4°F) or *severe* (less than 28°C or 82°F).

In the USA between 1979 and 1991, about 770 people have died each year from hypothermia.

Many standard oral medical thermometers measure down to 34.4°C (93.9°F). To diagnose and treat hypothermia, you need a thermometer going down to 15°C (59°F).

Hypothermia means that there is a mismatch between the amount of heat you generate, and the amount you lose. Hypothermia can be caused by environmental exposure, malnutrition and/or starvation, bacterial infection, severe hypothyroidism, liver failure, and various drugs (barbiturates, opiates, benzodiazepines, etc.).

The treatment of hypothermia can be very tricky. You can actually make the patient sicker, if you don't do things properly.

Suppose you directly apply heat (warm water or heating blankets) to the outside of the body. This can make things worse, by two different mechanisms.

First, the heat can open up blood vessels (which have been previously closed down by the cold) in the skin. Blood will flow into these newly opened blood vessels, and so the blood pressure can suddenly drop in your hypothermic patient.

Secondly, if you open up the closed blood vessels in their skin, cold blood can leave the skin and enter the central circulation, and make the core temperature drop suddenly. This is called 'afterdrop'.

In addition, there is always the possibility of burning the patient with an external heat source. In general, heat should be applied only to the chest and back.

Active Internal Rewarming is not a Do-It-Yourself technique and it requires special equipment.

An easy way is to get the patient to breathe warmed (42°C or 108°F), humidified oxygen via a face mask, or a tube down the trachea. This can warm up the patient by 1–2°C (1.6–3.2°F) per hour.

Probably the most efficient active internal technique is to warm up the blood externally, and then send it back into the body. This can increase core temperature as rapidly as 1–2°C (1.6–3.2°F) every 4 minutes. You normally use this extreme technique only in severe hypothermia.

No Bath -> Metabolism Adapts -> Temperature Up

The group of eight who did not take the cold bath seemed to have a so-called 'Metabolic Adaptation'. They felt very uncomfortable. Their bodies reacted to the cold by using more energy to speed up their metabolisms. This used up extra energy. Their BMR (Basal Metabolic Rate) increased by 75 per cent, and their rectal temperatures increased (+0.4 per cent), as their bodies responded to the cold. But while their core was hot, their skin was cold.

The nervous system entered a state of 'sympathetic' stimulation. Which meant that their endocrine systems released stress hormones (such as adrenaline and various steroids).

Cold Bath -> Hypothalamus Adapts -> Temperature Down

The three brave souls who did have the cold baths seemed to have a so-called 'Hypothalamic Adaptation'. Their bodies reacted to the cold by lowering the set temperature of their hypothalamuses. Their rectal temperatures decreased (−0.5 per cent).

They felt reasonably comfortable lying naked in a cold room.

Their BMRs did not change. There was no sympathetic stimulation of their nervous

  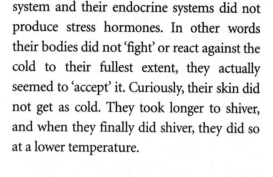

system and their endocrine systems did not produce stress hormones. In other words their bodies did not 'fight' or react against the cold to their fullest extent, they actually seemed to 'accept' it. Curiously, their skin did not get as cold. They took longer to shiver, and when they finally did shiver, they did so at a lower temperature.

The Test in The Arctic

Five days after the Nude Cold Tolerance Test, they all went to the Arctic. As part of the experiment they collected their urine samples for analysis on their return to the laboratory. (They froze the samples by putting them outside the tent.)

Arctic Results

The eight volunteers who did not have the nine cold baths fired their metabolic furnaces and raised their core temperatures, and also released more stress hormones.

The three volunteers who took the nine cold baths to reset their thermostat and drop their core temperatures did not make any stress hormones.

The results were pretty similar to the Nude Cold Tolerance Test.

This 'Chill-For-Winter-Comfort' effect lasted for about five weeks after taking their first cold bath; it covered their trip to the Arctic, and their return to Canada.

So if you can't afford to buy warm clothes or to pay big heating bills in winter, try pre-adapting your body to the cold. Start off with a cool shower every day, and gradually work your way down to a nice long, cool bath. If you want to enjoy winter, chill out!

Cold Bath → Greater Endurance

Now what about the cold baths improving the performance of certain athletes? According to Frank Marino and his colleagues at Charles Sturt University in Bathurst, a single cool bath can speed up the performance of an endurance athlete by 4 to 5 per cent!

Endurance athletes generate lots of heat in their muscles. This heat is carried by the blood to the skin, and is then dumped into the environment.

When you get too hot (hyperthermia), your performance goes down for a few reasons. First, your heart beats more rapidly for a given workload. Second, you send too much blood to the skin to dump heat — blood which could be used to power muscles. The hyperthermia can also retard your muscle performance by increasing your muscle temperature, and using up your energy source (glycogen).

A 'cooler' athlete has a greater temperature difference between the skin, and the environment, and so can dump heat faster. In addition, they start off cooler, so it takes longer for their bodies to 'saturate' with heat.

Cold Baths 'Reset' Thermostat for Athletes

But how do you make an athlete cooler? Refrigerated jackets (worn just prior to the race) have been tried, but they cool only the skin, and the effect is temporary. You really want to drop the core temperature. That's where cold baths come in again.

The Experiment

Marino worked with eight keen runners, three of whom were women. They were all to run as far as they could in 30 minutes on a treadmill, as it was gradually tilted to steeper angles. In addition the environment was hot (32°C or 90°F) and humid (60 per cent humidity). All the runners ran both with, and without, the benefit of a cool bath.

The runners prepared for the experiment by not having any vigorous exercise, alcohol or caffeine for two days before the experiment. They also had two practice runs on the treadmill, to get used to it, before the experiment.

On the day of the run, the volunteers turned up at the laboratory, and rested quietly for 15 minutes. They were weighed, had a blood sample taken, and had a rectal thermometer inserted. They then entered the water tank up to neck level. The water was gradually cooled from about 28.5°C (83°F) down to around 23°C (73°F). Seven of the volunteers were able to stay in the cool bath for 60 minutes, but one began to shiver vigorously and had to leave after 45 minutes. Within three minutes of leaving the bath, the athletes dried off, and began the treadmill run. Various measurements were taken every five minutes.

The Results

At the beginning of the run, the 'pre-cooled' runners had lowered their core temperature by about 0.7°C (1.1°F). Their heart rates were lower than the 'non-bath' runners, for the first 10 minutes only (and then the heart rates became the same). The rectal temperature of the runners was lower for the first 20 minutes only. The skin temperature was lower for the first 25 minutes only.

At every five-minute measuring point, the 'pre-cooled' runners had run further than the 'non-bath' runners. By 30 minutes, the 'pre-cooled' runners had run about 300 metres extra (7556 metres vs 7252 metres). That extra distance is about 4 to 5 per cent further, or almost another lap of the track.

The one major problem with the Chill-For-Endurance-Running Experiment was that the numbers were too small.

Athletes will try steroids, growth hormone, erythropoietin (and lots of other drugs with bad side effects) to improve their performance. Even if the statistics aren't strong, I reckon that some endurance athletes should try a cold bath or two in their training regime. Let's hope they don't get cold feet!

COLD SURGERY KILLS

If you're kept warm during an operation, your chances of surviving the operation are greater. It sounds pretty obvious, but it has taken surgeons and anaesthetists about a century and a half to realise this.

It's called 'hospital hypothermia' or 'iatrogenic hypothermia'. '*Iatro*' means 'doctor', and '*genic*' means 'caused by', so '*iatrogenic hypothermia*' means 'the doctors have caused a low body temperature'. In recovering, you generate up to 700 times more hormones of stress — and this can really give your heart a jolt.

A typical operating theatre runs at around 20°C. The surgical team are wrapped in gowns that keep them warm, they have layers underneath, and they're moving around.

But the patients wear nothing except thin sheets, and occasionally, a thin blanket. They are also motionless. When they have been cut open, chilled air will cool their organs. Because the patient has been anaesthetised, their thermostat (the hypothalamus) is asleep. So the blood vessels in their skin will not close down to keep the heat in their bodies, and they can't automatically shiver to keep warm.

In some cases, the first way you can tell that the patient is becoming hypothermic is that they keep on bleeding. The clotting factors in your blood (which stop you from bleeding) are switched off at low temperatures.

In fact, while you're anaesthetised, you turn from a warm-blooded animal to a cold-blooded animal, like a reptile or a fish.

During a typical operation, your core body temperature will drop by about 1.3°C (2.1°F).

The surgical people have long said that such a small drop in core body temperature really didn't matter. In fact, they said it might even protect you

COLD SURGERY KILLS continued

— by slowing down your metabolism, and by reducing your need for oxygen.

Opinions are one thing, but the only way to really find out is to *do the experiment*. Steven M. Frank (from the Johns Hopkins Medical Institutions in Baltimore) and his colleagues looked at 270 patients over the age of 60, who had been operated upon. One hundred and forty three of them had only thin cotton blankets during and after the surgery. But 127 patients were kept warm by special blankets that blasted tiny jets of warm air on them. The patients who had only the thin blankets, had their average core temperature 1.3°C (2.1°F) lower than the patients who had the full-body forced-air warming.

The doctors found three cases in which the cold patients suffered more 'heart events', ranging from *minor cardiac events*, such as a few ectopic beats that were not fatal, to serious events. They found that *major cardiac events* were 55 per cent less frequent in the warm patients than in the cold patients.

They also looked at *ventricular tachycardia*. This condition is quite serious. Ventricular tachycardia happened in 7.9 per cent of the cold patients and only 2.4 per cent of the warm patients.

The reason for these heart complications could be that once the anaethestic wears off the hypothalamus wakes up and tries to get the body temperature back to normal as rapidly as possible.

If you are only a little cooler than normal, the hypothalamus will kick a few systems into action. But when your core temperature is 2.5°C (4°F) below normal, the hypothalamus will increase enormously the levels of various stress hormones, by between 200 and 700 times. If somebody is already a little bit weak after an operation, a 700-times increase in stress hormones can have bad effects upon the heart.

With this knowledge, it seems pretty obvious that the surgeons and anaesthetists should switch over to keeping their patients warmer during surgery. It doesn't cost much, and it doesn't seem to cause any risks to the patient.

PROBLEMS WITH THE COLD BATH EXPERIMENTS

The Cold Bath Experiments need to be repeated with bigger numbers — say 40 volunteers. Three and eight test subjects are simply too small a group to run reliable statistics.

There's another problem with the Chill-For-Winter-Comfort Experiment. The three volunteers who took the cold baths weren't the same physically as the eight people who refused the cold baths. The three volunteers were leaner (less body fat), and fitter (could move more air in-and-out-of their lungs).

Even so, I'd bet a dollar that the effect is real, and that follow-up experiments will show that nine cold baths will *definitely* cut your hot water bill — and maybe your house heating bills too.

John Booth, Frank Marino and Jeffrey J. Ward, 'Improved running performance in hot humid conditions following whole body precooling', *Medicine and Science in Sports and Exercise*, Vol. 29, No. 7, July 1997, pp 943–949.

Wilson da Silva, 'If You Want Gold, Just Chill Out', *New Scientist*, No 2038, 13 July 1996, p 6.

Steven M. Frank, 'Perioperative Maintenance of Normothermia Reduces the Incidence of Morbid Cardiac Events', *The Journal of the American Medical Association*, Vol. 227, No. 14, 9 April 1997, pp 1127–1134.

M.W. Radomski and C. Boutelier, 'Hormone Response of Normal and Intermittent Cold-preadapted Humans to Continuous Cold', *Journal of Applied Physiology*, Vol. 53, 1982, pp 610–616.

R.B. Shukla, D.G. Kelly, E.J. Guiney and L. Daly, 'Association of Cold Weather With Testicular Torsion', *British Medical Journal*, Vol. 285, 20 November 1982, pp 1459–1460.

S. Sternberg, 'Warmth After Surgery Can Save Lives', *Science News*, Vol. 151, 12 April 1997, p 220.

'Pigeons landed a couple of scientists a Nobel Prize, because somebody bothered to discover the difference between Pigeon Poo and The Origins of The Universe.'

Proving that fact is stranger than fiction, Dr Karl Kruszelnicki launched his *New* Moments in Science series with this fabulous collection of stories about some recent awesome discoveries in science. How can sexual intercourse spoil your whole day (because you can't remember any of it)? What really killed Elvis Aron Presley (and why was he buried twice)? Why do men have nipples, why can the pill make a woman choose Mr Wrong, and why are there traffic jams in the middle of nowhere?

Dr Karl gets to the bottom of them all — and more — and then goes on to crack the big one: how *did* pigeon poo unlock the secret of the universe?

Yet more bizarre but true stories about the latest discoveries in science, by Australia's livewire guru of science, Dr Karl Kruszelnicki.

How can a string of big, fat lasers take the place of anti-missile missiles? What are engineers doing with a tuna called Charlie the Robofish? Do you know that bacteria live in strange, slimy skyscraper cities, and that some bacteria have a tiny electric motor?

Which star, 300,000 years ago, blew out a hole in our galaxy? Is it true that beer has fewer calories than skim milk? What should you do if you are trapped in quicksand?

Do you know that cordless phones can cause strokes? And that really soon (any time within the next few thousand years) we should flip into the next Ice Age?

The Fibromyalgia Coach

Strategies and support for your personal journey through FM

Pam Wright

The Fibromyalgia Coach

In memory of
Betty Hurrell

First Published in Great Britain in 2008

Contents

Foreword

I first met Pam a few years ago when she first attended for acupuncture treatment for her own fibromyalgia. Since then, I have been able to witness her remarkable progress with her initially debilitating disease. During the treatment sessions, Pam has often shared with me some of her 'coaching gems', which she not only uses for clients but also for her own benefit.

When Pam asked me if I would like to write a foreword for this book I was delighted and excited to get a preview of her 'trade secrets'.

I think that this book will be of great use not only to anyone with fibromyalgia, but also to anyone with other types of chronic pain conditions such as osteoarthritis or neuralgia. It gives concrete instructions on how to come to terms with the sense of loss that fibromyalgia causes and how to develop a positive attitude without surrendering to the disease. On this basis, Pam's coaching program guides the reader, step by step, on how to regain control over the symptoms, and how to deal with their own 'gremlins' which like to get in the way.

In my GP practice, almost on a daily basis I see patients getting stuck in the various pitfalls that Pam describes, and most of her case studies cause one or two of my own patients, who would benefit from her approach, to spring to mind. I even think that I ought to apply some of her tips on gremlins to myself.

I hope that I will be able to pass on some of Pam's advice in my short consultations. I certainly will warmly recommend her book to anyone who lives with fibromyalgia or other chronic pain.

Dr Gerhard Esser MD
Medical State Exam (Berlin)
General Practitioner with Special Interest in Acupuncture
Member of the British Medical Acupuncture Society
Canterbury, January 2008

Dr Esser is in General Practice in Whitstable, Kent

INTRODUCTION

Why This Book is for You

Are you, or is someone you know
- ○ trying to cope with the constant pain and fatigue of Fibromyalgia or a similar condition

- ○ looking for a no-nonsense, practical, positive approach to successfully managing and living life to the full with Fibromyalgia

- ○ wanting a holistic, realistic small-steps plan of action to improve the everyday reality of coping with a debilitating illness

- ○ needing support from someone who really knows what the problems are?

If you have answered 'yes' to any of these questions, look no further, this is the book for you. With tips, action points, and helpful information on ways to approach improving your present circumstances, as well as more individualised, reflective sections that move you forward with your life balance and future planning, this book will help to provide the solutions to any feelings of isolation, shock and worry about the future that you may be experiencing at this time.
The successful management of Fibromyalgia and its associated difficulties can be your reality. *The Fibromyalgia Coach* gives

you the opportunity to focus, re-think, re-frame, and take action to re-design your life. All the exercises, questions and suggestions have been tried and tested. They really work. Although long-term results will not happen overnight, great changes *will* happen through small manageable steps, according to how much time and energy you invest in yourself. The reward will be your own well-balanced, stress-managed, interesting, fulfilling life, in which you are in control and able to make choices that keep you as fit and as well as possible.

To become a successful manager of your life with Fibromyalgia you need:
   ❖ time (when you're ill you have time, even though it may be unwelcome)
   ❖ courage and self-belief (don't worry, you'll find this through our work together)
   ❖ determination (we will be working on finding the right motivation to keep you moving forward)
   ❖ a sense of humour (if you've lost it, you will find it again)
   ❖ a willingness to think differently, and to be open to new ways of looking at your life.

Feeling unsure?
Remember, this is only a book; you can pick it up and put it down as you feel able. Nobody will be checking up on you - you are in charge of you. If you decide not to take the contents seriously, then nothing changes. But if you work through and apply the coaching to your own particular reality you will be taking great steps to regaining control of your life. It's your choice!

Everything I tell you about has already worked for others, including me. I continue to experience and work on all the thought processes, practical steps and pacing skills within my own successful management of Fibromyalgia. I therefore know many of the physical challenges and mental difficulties that you may be dealing with. Everyone's life is their own, and as an experienced personal coach I can ask you the right questions to enable you to search for and find the right solutions for your own circumstances. So, let me work with you, guide you and support you through the positive changes to come. There's a whole new world out there just waiting for you. You've already been thinking about taking control and working positively, otherwise you wouldn't be reading this. Now is the time to move forward.

FIBROMYALGIA
A Hidden Disability
Fibromyalgia (FM) is one of a group of conditions described as Central Sensitivity Syndromes. It is a muscular-skeletal pain problem that affects the lives of millions of people throughout the world. Its main symptoms are widespread chronic muscle pain and severe fatigue. People who have FM show no outward physical signs of illness (unless they need to use a wheelchair at times). Unlike some rheumatic diseases in which joints become mis-shapen and the lack of mobility is obvious, people with Fibromyalgia look unaffected

and often have difficulty explaining the reality of their symptoms.

At present, FM is diagnosed by eliminating all the other illnesses and conditions that the symptoms may originally suggest. Tender points in the body are identified and pain levels monitored. The basic, but constantly changing criteria for diagnosis at this point depends on a person having chronic pain for at least three months, and the presence of 11 out of a possible 18 active tender points in the body. Depression, sleeplessness, cognitive dysfunction, irritable bowel syndrome, restless legs syndrome and many other associated conditions all add up to a diagnosis of FM.

At this time, there is no effective standard cure and, because it is not seen as an acute condition, the available medical care and expertise is patchy and variable in its patient support. Research is under-funded and unco-ordinated in its approach. Fibromyalgia, itself, cannot be described in a medical sense as a life threatening condition but, for those who have to live with the symptoms, it is life changing, and if we do nothing to help ourselves our quality of life can become so severely impaired as to become life threatening in every practical sense.

LIFE COACHING
The Positive Way Forward
Life Coaching helps you to successfully manage life changing situations and health issues. It is a practical, holistic process that works on moving you forward from where you are now in your life, to where you want and need to be in order to live life

to the full. Coaching helps you to use, effectively, all your inner untapped human strength. It enables you to become more positively self-aware and to exercise choice, control and freedom in a way that is right for you. There is no one-fit solution and there is no magic formula for success. Each person has a different reality and perception of life, which has evolved through their past experience, present reality, and their hopes and expectations for the future. Working, systematically, through questions aimed at bringing your mind, body and spirit into balance, and turning thoughts into positive action, coaching lifts the lid off your potential and enables you to live life, rather than continuing to let life live you!

In this book, you will be helped to identify your practical, emotional and spiritual needs and then to match your ideas and practical plans to your life values. You will learn how to banish your self-doubt, improve your self-esteem, cultivate your self-reliance and self-image, prioritise your actions, discover what motivates you and make that movement forward by taking action at a pace that is right for you.

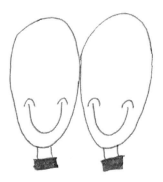

OUR FIBROMYALGIA COACHING JOURNEY TOGETHER

Although complete freedom from FM seems to be an unrealistic goal, at present, it is worth keeping that ideal in mind. We will be setting our sights on minimising the negative affect that FM has on life to such an extent that we no longer feel held back by the condition. We are about to undertake that journey together. It is your journey, but I am with you as a guide, supporter, non-critical ally, motivator and challenger to help you to regain your life balance and to live life to the full. You will learn to use all your skills and ability as a person – lots of which you may not have discovered yet.

You have what it takes to be successful.
It's not easy (particularly in the first few months of realisation) to accept that Fibromyalgia is now part of your life, but don't despair; it is possible to successfully manage your symptoms so that you feel that the condition does not stop you being the person you really are or want to be.

Although, at this time, there is no single magic pill for the relief of all symptoms, there is a huge bank of helpful information, which has come from other successful Fibro-fighters who have been through the uncertainty and fear you may be experiencing now. Using their experience, ideas and tips and adding your own special understanding of what helps, you, too, will be able to get on with your life, in spite of having FM, and regain control by designing the life you need and want.

Find your personal inner strength.
Have you ever wondered what it is about some people that they apparently sail through difficulties and keep smiling? The good news is that such inner strength is in all of us. It just needs a chance to come forward and to be given a little encouragement by the one person who needs it most, right now – *you.*

My promise to you is that if you give yourself a serious chance to manage your life with Fibromyalgia by giving yourself time and space to make changes, by the end of this book you will have significantly lightened your present load. You will have made many positive changes, and you will also be well on the way to achieving the optimum successful management of your life.

I shall be helping you to:
- ○ understand more about yourself by finding out what really makes you tick
- ○ recognise, overhaul and change if necessary, your present coping strategies for when things are not as you would wish them to be
- ○ find different ways of identifying ways forward which are linked to your likes, dislikes, thoughts, feelings and possessions that you value
- ○ cultivate a realistically positive outlook on life
- ○ feel less alone with your problems
- ○ improve your self-belief, self-esteem, self-reliance and self-image.

You need to be:
- ✓ willing to look in detail at how you function
- ✓ honest with yourself and in your answers to the questions posed
- ✓ prepared to try out new strategies and ideas more than once if necessary
- ✓ aware that although your life is unique, there is always someone somewhere who can help you move forward positively.

What have you got to lose?
Let's work on it together!

HOW *The Fibromyalgia Coach* WORKS

This book concentrates on the here and now. It helps you to get to grips with what is happening to you on a daily basis and to prioritise your needs. It enables you to communicate effectively with friends and family. It gives you the tools and confidence to approach employers, doctors and other medical practitioners, and benefit agencies, all of whom deal with factual information in their planning. It also helps you to find strategies to bring everyone you meet to a greater understanding of what is happening to you.

Most of all, it helps you to realise that your life is still your own and that you are still the same person as you were, even though you are now being challenged by symptoms and situations that are largely beyond your control. It enables you to move forward at a pace that you can manage, which may vary from day to day or week to week, so that you feel once again that you are in control of your life rather than feeling that life controls you.

There are **10 Success Skills** to learn, practise and use in everything you do and say. Sounds weird or hard? Don't worry, they are presented in bite sized chunks with lots of practical examples of how to use them. They focus on what really matters when you are ill and are very much based in common sense and effective communication, which will also help you when you are feeling much improved. They are designed to cut through the fog of indecision, worry, anxiety and fear that is so common whenever we are stopped in our tracks by any sort of illness. Fibromyalgia can be even more difficult because of the 'mystery' of how it affects us.

There are **20 Action Points** that build up your self-knowledge, self-confidence and self-esteem so that you really know, and can communicate with yourself and with others, exactly what you need at this time. Some of the Action Points  involve compiling charts and keeping records. Copies of these charts and other useful information are to be found within the text but also at the end of the book, so that you can make copies for your own use and continue to put everything into practice until such time as you realise you can do without the extra support. I am not suggesting that your life should be ruled by bits of paper forever, only for as long as they are useful to you. The nature of FM is that when you get through the worst of it and start enjoying life again, there may be times when a quick revision of what you learn here will boost you to move forward once again. Remember, you will never again be in exactly the same position as you are at the moment.

Together with the Success Skills, these Action Points enable you, with increased positive self-awareness, to build up a plan based on achievable, realistic goals that are in line with what you feel really matters to you in life.

Although it is possible to gain some benefit by using any section in isolation, the most effective, significant changes will happen when you work, systematically, through the book. By thinking seriously and honestly about all the issues, then completing the exercises and spending time using the information to plan your changes, you will grow through the experience and move ahead with confidence, safe in the knowledge that you have built up the right life formula for your own particular needs.

There is no time limit involved, particularly as Fibromyalgia is not a time-related condition. You decide to take on as much or as little as you are able to, according to your reality. But my guess is that once you have even the smallest success, you'll want to keep going. Good luck! **Your successful management of life with Fibromyalgia starts here!**

FIRST THINGS FIRST

1. Buy a hard-backed book in which to write or draw. You can then use it in a reclining position or in any chair, without the need for a table.

2. Buy a pen with a fat, round or triangle-shaped shaft, which is comfortable to hold and needs very little effort for the ink to flow.

3. Set aside a time of day when you can be awake, quiet and alone. This may take some organising, but if you are serious about helping yourself, it is essential.

4. Choose a comfortable position in which you can read and write: you could be propped up on a bed, sitting on a sofa, at a table, or on any chair. Have a variety of seating which will suit your needs according to how you feel at any particular time.

5. Buy a kitchen timer, or use a digital watch or mobile phone as a timed alarm. By trial and error you will know how long you are able to sit in one place before having to move. Use the timer to remind you to get up or move and stretch, or to stop working and change activity.

6. Make a snack or have a drink ready for yourself for when you finish your session, and have mineral water near you while you are working.

1

Acknowledging the Past

Where have you been?
Each one of us is a unique being. Although it is possible to generalise about type, character traits, family background, childhood and life experience, the only person who truly knows what your life feels like, is *you*. Your past, your personal relationships and your expectation of what is possible for you are all coloured by your perception of life. You perceive things based on your own understanding. Your own understanding comes from what you know as fact and what you believe to be true. This explains why, when a number of people are witness to the same incident or event, they will all give differing descriptions. Similarly, a childhood memory will evoke different aspects, thoughts, feelings and reactions in a discussion between siblings who were all present at the time and apparently shared the same experience.

You are entitled to hold your own view of events in your past.
Few people come through life without being affected at some time by hurt, anger, grief or loss, separation, and various types of abuse or neglect. We all build up our own stock of reasons why we are the way we are. You know, already, what you have lived through to get where you are today. Many of us underestimate just how much personal courage and fortitude it has taken to manage when things are tough. We just get on with it because we know nothing different. But *now* is exactly the right time for you to recognise and acknowledge your triumphs, however small they might seem to you. It matters in your quest for better health that you know inside that you are already a survivor. Your inner spirit helped you get through

things before and it will move you forward once again, if you give it the opportunity.

Counselling or coaching? What is right for you *now?*
If you are, as yet, unable to recognise or understand how much your past has shaped your life's path up to the present time, the first questions you need to ask yourself are:

- ❖ how might learning more about myself, by looking in depth at the past, be of help in my quest to move forward
- ❖ do I understand how events in earlier years have led me to where I am now
- ❖ will delving more deeply into the past make any positive difference to my present and future?

If you are unsure of the answers, it may be that the first action for you to consider is whether you need to have some counselling before becoming involved in the coaching process. An initial no-obligation conversation with either a life coach or a counsellor (or both), about your feelings, may be helpful to you in choosing the first step of your route forward. Coaches and counsellors, alike, are bound by strict ethical codes that involve offering to refer people on to other professional services if their own service appears to be inappropriate for that person at that time.

Counselling and coaching are different.
Counselling helps you to understand where you have come from and how your reactions and decisions may have been affected

by events or relationships. It helps you to come to terms with the past, and to work through feelings that surface in the process, in such a way as to enable you to move forward with greater awareness and understanding of your roots or previous path. You may need a counsellor to help you come through some deep-seated anger or frustration about how people behave towards you. Unhappiness, bitterness and blame can sometimes lead to you feeling like a victim, and indeed, by feeling like one, you can fulfil your own prophesy by actually becoming one. Finding out where that idea originated may be illuminating and free you up to feel more able to look at your own behaviour in a different light.

ROBERTA'S STORY: counselling was needed before coaching could help.

Roberta's body language, when she sat down in front of me, was angry from the start. My intuition told me that I was on trial, quite an unusual feeling for a coach whose role is to enable people. Her face and manner indicated she was already expecting something less than positive from me; her defensiveness was almost palpable. I explained that I would be asking questions to which there were no right or wrong answers, and that it was part of the process of establishing a good understanding between us, as coach and client. I was unprepared for the tirade that ensued. It is still etched in my memory after many years of coaching clients. Roberta asked me what I thought of her appearance, how old did I think she was and what sort of person. Without waiting for a response, she explained that everyone she had ever met judged her, her workmates were all against her, nobody cared about her aches and pains and she felt completely let down by everybody. Life was completely unfair, she hadn't asked for such burdens, and even her family treated her poorly: nothing at home ever went right, either.

Roberta had arrived at the conclusion that she wanted to change her life but was looking for the wave of a magic wand to create that change. It was clear she was not ready to take any responsibility for what was happening in her life and, therefore, was not yet able to be an equal partner in the coaching process. We parted company, amicably, after I suggested that she would be better off talking to a counsellor or psychotherapist in order to get to the root of her feelings before she attempted to have coaching.

Coaching is based on empowerment and taking positive action to change oneself. Some counsellors also have coaching qualifications, which can be useful, but many people decide to draw a line under the past with the end of the counselling experience and start afresh with a coach.

Coaching acknowledges the past in less depth and focuses your attention on your future. Past mistakes and scenarios are discussed only for you to use as learning opportunities and clues for making decisions on what to change, discard or rediscover. Coaching is also based on practical action.

HELEN'S STORY: coaching was exactly right for her.

Helen was a bright, striking young woman whose high hopes for involvement in performance art were dashed by the onset of Fibromyalgia. Following a difficult childhood, she had shown huge inner strength by studying to a high level in a number of areas, including dance, drama and hairdressing. She contracted Fibromyalgia after a car crash. The ongoing pain and fatigue gave her no choice but to give up her home and move back in with her mother, with whom she had been living for 2 years before coming to see me. During that time, Helen had not travelled anywhere without her mother for fear of being overcome with fatigue and not being able to get home again. She now sensed that her mother was beginning to resent this dependence, and Helen, herself, after once being completely independent, was feeling more of a burden than she wanted to be. In addition, her boyfriend, though caring and considerate, did not understand the ups and downs of the symptoms that she dealt with on a daily basis.

During her first session, Helen sat with her feet tucked up underneath her, hugging her knees as she told me the nature of her present reality. She identified that she had lost confidence in herself, was worried about the future, about people's expectations of her, and about how all of this was affecting her mood and her symptoms.

Helen needed very few sessions of coaching. The second time she visited me, she travelled alone for the first time without any difficulty (we had worked out safety strategies previously). Within a few months Helen had moved back into a place of her own and was coping with great confidence and enthusiasm. She had redefined and improved her relationship with her boyfriend in such a way as to take the pressure off both of them. She had been able to explain her own needs to her mother without seeming ungrateful or feeling selfish. (She said that her mother was enjoying her own new-found freedom again.) Having recognised that she could use her talents in many different ways, as well as manage her Fibromyalgia successfully, Helen took the coaching strategies and flew with them. The last I heard was that she is running performance art workshops for children, in London, and is enjoying life with the minimum of interference from FM.

BETTY'S STORY: she used both coaching and counselling to great effect.

Betty was already 77 when she came to me. She had had a number of illnesses since retiring from teaching, but it was Fibromyalgia that she found totally debilitating, irritating and frustrating. She had always been a person who got things done but, as she said to me, 'My get up and go has got up and gone.' Betty was not a stay-at-home type. She had passed her advanced driving test at the age of 75 and always felt that if she had nothing to show for the day she had done nothing, and that made her cross. She told me she wanted to get her life back and she was determined to do so.

  Betty's story is inspirational. She worked on small steps of self-appreciation (which she thought very silly at first!). Soon, she began to realise just how much she was appreciated by everyone and was able to relax in her own space by *being* rather than *doing* all the time. She still continued with photography, using a computer, doing woodwork, card-making, T'ai chi, gardening and walking when she was well enough to do those things, but she was able to accept that it was okay to have 'time out' as well.

There is always a reason why people feel driven. It varies, of course, from individual to individual, but often it is wrapped up with feelings of self-worth, or the lack of it. Having dealt with her on-going life in the present, Betty recognised that most of her anxieties, about having to prove she was good enough, came from her childhood so many decades earlier. She decided that she had a particular childhood issue she wished to talk to someone about and began work with a counsellor, while still working with me. She did not want to take a break from coaching because the small, positive steps in everyday management of her life was having a hugely beneficial effect, and she was smiling and laughing again in spite of her symptoms. She sometimes shared some of the information she had learnt about herself with me in our coaching sessions, and I was able to strengthen her new understanding of herself by pointing out how she had used her experience to benefit others, and continued to contribute so much by her very positive attitude.

Betty never stopped organising her life and doing new things. Even in her last illness she was instructing the hospice doctors in the way that Fibromyalgia affects people. She managed to get them to recognise the 'special needs' understanding required about the extra sensitivity to drugs and pain relief for anyone who was dealing with FM alongside more obvious and trackable illnesses. Betty (of whom you will hear more later) remains an inspiration to all who knew her, including me.

Looking at the examples of Roberta, Helen and Betty, perhaps you can now confidently ask yourself the question: have I come to terms with my past and am I ready to use appropriate information in a positive way to help me design a better future for myself?

If the answer is 'yes' – great, let's get moving!

SUCCESS SKILL 1
Noticing and Zapping Your Negative Gremlins

A gremlin, in this context, is a negative thought or a self-limiting belief that stops you daring to try. You may have no difficulty identifying your gremlins, but, if you do, they can be recognised as quite simple thoughts that affect you negatively, making  you avoid or put off what you would really like to do. Zapping your gremlins means recognising these thoughts so as to be able to defuse more readily, the feelings that come with them, such as fear or loss of control.

Here are some common gremlins. Are these yours too, or do you have different ones?
I've never done it before.
I don't know where to start so it's not worth trying.
What will people think? How will I cope with the comments?
I couldn't do that on my own. I'm shy.
I haven't got time. I haven't got confidence, training or experience.

Sometimes, *people* can form part of your 'gremlin community'. Even a negative look or mildly dismissive comment, especially from someone who matters to you, can have a devastating effect on whether your gremlins rise up in your mind and take control. If you know someone whose manner, or usual reaction to any form of positive change you try to make, results in you doubting yourself completely or leaves you drained of energy, you need to know what is happening and learn to deal with such a situation differently. Don't worry, as time goes on we will be working on eliminating the power your gremlins have over you so that, if they return (as they do for all of us when we least expect them), you are able to recognise, rethink and repel them or kick them into touch, enabling your self-belief to remain strong.

When life is difficult and everything you try to do takes a huge effort because of pain, fatigue, depression or the frustration of feeling you are not making headway, it is easy to let inner negative thoughts take over. It is also easy to believe that every day is the same and to forget that your symptoms vary in their intensity from day to day, hour to hour or even minute to minute within each day. The truth is that by learning more about how you are affected, by focusing on yourself, your triggers, your patterns of reaction and fatigue, you will be able to see windows of opportunity through which you can start moving forward. Even if it feels that only the tiniest amount of light and air are coming through that window, at the moment, it is, nevertheless, an opening to the wider world and your much brighter future. We are aiming to be able to fling open wide that window of opportunity by the time you have worked through this book.

Turn the negative into positive.
Recognising the difference between what is an unhelpful negative thought and a positively helpful mental reaction to a problem (however small) means that you can choose to zap the negative and turn the positive into real action. Everything we do involves choice. There are many times in life when we feel we have no choice, but the reality is we choose to stay put, do nothing or do something for whatever reason we believe at the time. Illness stopping us in our tracks is one experience when we have very little choice but to look around and to take stock. Even then there is choice, either to become the chronic sufferer and to make that your new 'career', or to come to terms with what is happening to you, accept it and take control by working on ways to minimise the difficulties and get your life back – which is what we are embarking on together, now.

Know yourself.

Ask yourself the following questions, which are designed to get you thinking about your present and past coping strategies. Think about how you were before FM and also how you are now. Apart from the obvious symptoms, is there a real difference or not?

❖ When faced with a decision or the need to get something done, what do you say to yourself? *oh god! another ting! Cmon get this done quickly*
❖ Do you find the energy to do something for someone else, but not for yourself? *Yes!*
❖ Is it normal for you to use words like *but, if only, I can't,* or *I don't like to be a nuisance and ask for help? (Could they be your gremlins?)*
❖ Is your reaction directly linked to how well or how badly you feel, physically? *Yes*

It is natural that you may believe that your 'normal' has changed since FM joined your life, but an honesty check is needed here: if you have always tended to put things off, even when you were well (another gremlin?), then you need to accept that as a personal habit having little to do with your present circumstances. As you progress through these exercises, you may choose to decide that you do not wish to continue to be a procrastinator. If, however, you know that any hesitation is directly associated with your FM, wait until you feel less painful or fatigued, and ask yourself the question again. Either way, learning to pace yourself and make decisions that are in the best interests of your future good health is part of designing the new you. This is what we are working on, even now.

Illness can bring about a loss of self-confidence. This is made worse if your self-confidence was not one of your strong points to begin with. It can be just as difficult to cope with if you have always been a dynamo and are now trying to deal with the frustrations of being unable to maintain all that you feel you 'should' or 'ought' to be doing. Whatever your reality, as we work through the following chapters together, your self-confidence, motivation, self-reliance and self-preservation will rise to great heights. *Choosing* to put off something will then be *because you have made a confident and informed decision* that is in line with your life values, not because you are unsure of yourself, don't know where to start or feel compelled, for whatever reason, to continue even if you are aware that it is doing you no good.

When JILL came to me for coaching she was totally demoralised by her circumstances. She had been a dynamic, creative organiser with an artistic flare that embraced a number of skills, including embroidery and calligraphy, before Fibromyalgia and other health difficulties unexpectedly took the fun out of her life and slowed it down. She now spent all her energy just trying to look after herself. From our conversation, I soon realised that Jill's perfectionism had turned into a gremlin for her, and she was seriously in need of rekindling the great feelings she formerly got from her creative art work whenever she felt like it. Since becoming ill, she had occasionally thought that she would like to get back to her calligraphy and had looked longingly at her art table, which was covered in books and papers that (her gremlins told her) needed to be sorted and tidied away before she could feel free enough to dare to get the inks out again. But by then she would have run out of energy and the moment would be lost, so what was the use? And anyway, her hands were playing up with intermittent pain and she wouldn't be able to work for as long, or to the standard she used to achieve, so again, what was the use?

We focused on the minimum amount of change required to help set up the right conditions for Jill to feel more positive about returning to her artwork. She had actually thought she was doing nothing at all creative, but soon realised that in her art play with her grandson she was continually using her creative talents, albeit in a different way. Her grandson was completely captivated by the artwork he did with Jill, and she loved seeing his face as he experienced the same thrill of creativity as she, herself, always had.

Looking at her own artistic needs from a new perspective and deciding that lots of the thoughts were gremlins to be dealt with, Jill took the following actions:

- she dumped the piles of paper and books on the floor in order to clear the space on her art table
- she put her art equipment ready on the table
- she made sure that her favourite chair was in place and ready for her at any time.

The positive effect on her morale was almost immediate. She realised that the next time she felt able and thought she'd like to do some artwork, there would no longer be a physical barrier to getting started. Just handling the equipment again, knowing that that possibility now existed, gave her such a buzz that she resolved to set about buying more storage space for the papers and books that had apparently kept her from moving forward for so long. Within a few months, her art area was clear of the extra clutter on the floor: she had sorted through it, bit by bit, at a pace she could manage. She had also built in short but totally satisfying creative times during her week. Jill had learnt to recognise her gremlins and zap them. She tells me that whenever she realises they are creeping back, she deals with them swiftly, and each time it becomes easier and more satisfying to be so empowered.

ACTION POINT 1
Thinking back

We all need to acknowledge where we have been and how we have managed, if we are to make good choices for ourselves. The following questions and statements have no right or wrong answers. They are designed to help you identify your feelings in good times and to give yourself credit for coming through difficult times. Your answers may also give you a clue as to what motivates you and what you wish to avoid in the future. They will also help you to get through the present, particularly if, like Jill, Helen and

Betty, you feel really stuck and frustrated right now, and want to take back control of your life.

Don't spend a long time thinking about the answers. Write down what comes to mind quickly and if nothing comes, leave a blank. Sometimes, answers will pop into your head further on in our journey together and you'll wonder why they had not been obvious before. That's absolutely normal, and nothing to worry about.

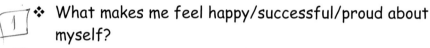

- ❖ What makes me feel happy/successful/proud about myself?
- ❖ When was I at my most relaxed and happy?
- ❖ What has always made me feel good?
- ❖ What has been happening when I have really felt comfortable with life?
- ❖ What's the biggest thing I have had to overcome in my life up to now?
- ❖ What has not been so successful, and what have I learnt from this?
- ❖ How has my attitude to people changed in recent years?
- ❖ Do I think mostly about the past, the present or the future?
- ❖ What is the most important thing in my life at the moment?
- ❖ 5 positive things about me are:

Looking at your answers, how have you described yourself? Are you complimentary about what you have achieved so far? Or does the very word 'achieve' send you into a spiral of self-doubt and negativity? In the next chapter we'll be working on how to put everything in a more positive light, but first, let's have a look at what we have covered already.

Review

You have been acknowledging your past by:

- o recognising that nobody other than you knows what your life is like
- o accepting that it is all right to hold your own view of your life events
- o considering if you need counselling, coaching or a mixture of both.

You have become acquainted with the idea that everyone has 'gremlins' that cause lack of confidence. You are preparing to Recognise and Zap your Gremlins (Success Skill 1) by:

- o understanding how you normally think when faced with a problem
- o thinking back to better times to identify good feelings and enjoyment
- o starting to give yourself credit for all the things you have done up to now.

That's a great start! Don't forget to take breaks and to give yourself a 'change of scene' from time to time. Even sitting in a different chair provides a new perspective on your surroundings while you do this.

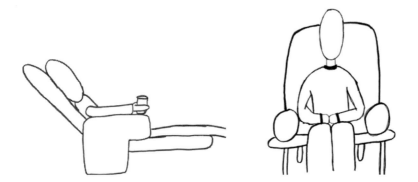

2
Choosing a Different Way

We've briefly looked at the bigger picture of your life and I hope you are already beginning to notice and give yourself credit for just how much you have achieved in the past. We have also identified that everyone has personal thoughts that can hinder them in the pursuit of their hopes and dreams, but what about dealing with the unavoidable everyday negativity that is the reality of coping with an open-ended, debilitating illness like Fibromyalgia?

This chapter contains some of the most empowering skills available to anyone. Take it seriously, but have fun with it too. Try to notice how the difference in your attitude to yourself and others improves as a result.

SUCCESS SKILL 2
The Flipover: there is always another viewpoint

If you see a half-filled glass of water, do you describe it as half-empty or half-full? Would you say 'there's still some left in the glass', 'it's almost gone' or 'great, here's an opportunity to quench my thirst'?

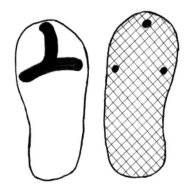

Learning to reframe, or 'flipover', negative thoughts is a skill that leads to a more positive outlook, which in turn lifts the spirit and provides hope and the motivation to take greater control in your life. The Flipover skill takes practice, especially if you have only been used to noticing the negative in conversations, situations, the weather, and of course your FM. But half the battle is awareness, and once you start to notice what is happening, then you can choose to make a difference in the way you think. It can be used in any situation, anywhere at any time.

- ✓ Banish all the negative thoughts that crowd into your mind by replacing them with simple statements that help you to see your perceived difficulties from a more creative angle.
- ✓ Think what you *can* do rather than what you *can't* do.
- ✓ Be aware that having to behave and act differently, at the moment, because of your condition, may be giving you an opportunity to do something that you've never done before or have been wishing you had time to do for years.

I used the Flipover when I was first ill with FM to help me cope with the very real sense of loss.

I was in total shock about being physically unable to do my job, lift a saucepan or make a bed (that's if I managed to struggle out of it at all on some days). Unlike many people, I had always liked the bustle of Monday morning in the school where I had worked for some time. The fact that I was now stuck at home while all that activity was continuing without me and all my friends were at work, caused me real grief and despair, particularly as, at that time, nobody could find out what was wrong with me. After a few weeks of an emotional downward spiral, I decided that some form of routine was needed if I was to have any hope of staying out of the pit of despair that seemed to be beckoning.

I used the Flipover skill by thinking that although I might not be able to drive, walk far or do my job, what I *could* do, now, was to be spontaneous with every day that dawned. It was autumn, with days alternating between bright sunshine, with a hint of crispness underfoot, and dreary damp mist that made every muscle feel miserably painful. I decided that on Monday mornings, whenever I could and if the weather was fine, I would make the effort to walk the short distance to the beach. If the weather was bad, I would give myself the luxury of taking my breakfast back to bed and listening to the radio in relaxed comfort until I felt able to get up.

As the weeks went on, I began to marvel at the fact that, although I was far from well, I was lucky enough to be able to watch the seabirds feeding and the oyster fishermen dredging just offshore, even if only for a short time before my symptoms kicked in and I had to return home. I began to enjoy the feeling of snuggling down again in bed, when the weather was bad, without the awful sense of guilt and loss that had been with me at the beginning of my FM journey. After all, at all other times I was doing my utmost to find out what was wrong, and there is a lot of waiting time between medical test appointments and results so it was all out of my hands until then. I had used the Flipover to enable myself to recognise that although there was a negative side to my situation, there were also positives to enjoy. I had never before allowed myself to give time to just standing and looking around me, and I could have counted on one hand the number of times I had allowed myself to stay in bed on a Monday morning without having a high temperature!

The Flipover had enabled me to realise that acceptance of my situation had given me different things to appreciate. It was the start of my truly positive thinking process.

ACTION POINT 2
Positive questions to ask yourself when faced with a problem

The Flipover skill helps a great deal when finding solutions to the practicalities of daily living with a debilitating health condition. In Action Point 2, use the Flipover skill to help you find a positive spin on everything you are trying to do. Within the questions, I have given some examples of Flipover thinking that may help as you go along.

❖ **How can I get this done?**
Write down all the possibilities, even the crankiest of ideas. Think 'round the problem', 'over the problem', 'through the problem', and 'under the problem'. You will probably find that the problem, itself, becomes less of a depressing issue as you think creatively and use the new skills you are acquiring to find the right solution for yourself.
Also ask yourself, 'Do I really have to do this at all?' and 'How important, in the wider scheme of things, is this

problem?' (You are already using the Flipover skill here to good effect.)

❖ What will happen if I don't do this?

Be honest with your answer. You might find that you bring on more things to do and your stress levels rise, or you might realise that it's not as important to complete as you first thought.

If you are sure you want to get this problem sorted out, move forward to the next question.

❖ What can I do myself?

Be positive, realistic and think creatively. Consider the best time of day for you to undertake any task, and make sure you think about how much energy it will take out of you, or if you need to do it in short bursts.

(The Flipover skill statement here would start with, 'I will choose to do specific tasks that I feel I can manage at times during which I am at my present best.')

❖ Who can I ask to help me?

Avoid negative people, particularly people who drain you of what little energy you may have at present. (The Flipover here is, 'I will choose only positive people to help.' And if you feel you need to justify your actions to anyone in particular, explain that you feel they have enough on their own plate at the moment. Being a negative person, he or she will most likely agree.)

❖ **What do I need to make this happen?**
Identify everything you need, depending on what you want to do. The list should include all the things you will use (e.g. phone, computer, paper, envelope, stamps), as well as need (e.g. peace, space, one of your better days – and actually write down how you need to be feeling on the day when you will be well enough to take some action on this issue). You may even need co-operation from a friend or family member, and that needs to be sorted out ahead of time also.

❖ **What do I need to do first?**
What's most important to start with? Look at all your writing and choose which task (however small) will set you on the road to sorting out this particular problem.
Use the Flipover skill, constantly, by reminding yourself that although you are not as well as you would wish to be, as yet, you are still managing to get things done that are important to you.

❖ **What do I need to do next?**
Decide on one step at a time. Don't rush. Enlist help (only from people who are reliable and will help you with enthusiasm and a smile) or take each task as it comes when you are ready to manage it.

❖ **When will I do this?**
Name a day or the next opportunity, but don't go into, 'I wish I could do it now' thoughts. Instead, Flipover the idea and say, 'I will do some part of this on the very next day

that I am able to manage.' You may find that is sooner than you think.

❖ **What will I feel like when I have achieved this?**
Picture yourself feeling great, happy, fulfilled and positive. After all, you may be having a difficult time with FM but you are still you, and your personality is still intact. Things could be worse (now *there's* a Flipover remark!).

Review of Success Skill 1: Zapping Your Gremlins
In Action Point 2 we worked on positive questions to ask yourself when faced with a problem. One of them referred to asking for help from somebody else, if necessary. Are you, like so many independent people, so used to getting on with things that you have forgotten the art of asking for help for yourself? Perhaps you are always helping others and give the impression that you are never in need of help yourself. Thoughts such as, 'I'm ok', 'I can manage', when you are actually in real difficulty, give way to 'I don't like to be a nuisance' or 'they're bound to be too busy', and are actually the gremlins of negative self-belief rather than true fact. (Mind you, if someone has always told you they are too busy, avoid *them* at all costs – we are out to improve your self-esteem and self-worth not squash it further!)

      To zap this particular gremlin, think for a minute: has someone *actually told you* that you are a nuisance? If so, don't ask them for help, ask someone else. If not, feeling that you are 'being a nuisance' is to do with what is going on in your head rather than true fact, and it's time to change this thought so that it no longer stops you from doing what is good for your own health and wellbeing.

Most people like to feel that they can be helpful to others, especially if they know they are making a positive difference. There's nothing to lose and everything to gain by daring to ask for help. The worst that could happen is that they say 'no', in which case you ask someone else, or, you could be happily surprised by the positive reaction to your request. And into the bargain, you have started recognising thoughts that hold you back unnecessarily, and are taking steps to deal with them. Gremlin Zapping (Success Skill 1) rules, okay! Keep practising!

Making friends is a social skill that some people find easier than others. We all have different experiences of meeting people who come in and out of our lives quickly or intermittently, while others stay around for years; acquaintances become friends, and friends become good friends or just 'fair weather' friends, according to their life agenda and yours. Friends help to provide us with our sense of identity. The best are kind and considerate. They go the extra mile for you when you least expect it, and there is an unconditional sense of shared history that makes the friendship strong.

SUCCESS SKILL 3
Be Your Own Best Friend

In view of what I have just said about friends and friendship, it may seem strange to you that Success Skill 3 is to learn to 'Be your own best friend at all times'. It is something that we all forget about in the quest to be popular, loved or surrounded by people.

Contrary to what the media reports on a daily basis, many people still follow the life value of behaving towards others as you would wish them to behave towards you, but the expectation of hoping to receive similar kindness, consideration, thought and care in return, can turn some people into disappointed and bitter souls. They wait, in vain, in the belief that their own happiness is dependent on others 'behaving well' towards them. This is not the case. When you have a vibrant sense of self-worth, the need to receive constant affirmation from others fades, and you make decisions according to what you feel is right for you. You choose to be with people who provide what you need and need what you can offer them. That is not the same as being selfish, and selfishness is not the same as self-preservation, which is what you need to be aware of now. Self-preservation, for me, only really became an issue through my own ill-health. Suddenly, it dawned on me that until I thought I was worth preserving I could not expect anyone else to believe that I was. I needed to be more comfortable with who I am and to be nicer to myself, particularly as, like many people, I was too ready to accept responsibility or blame in the interests of a peaceful (but uncomfortable) existence.

A selfish attitude is one in which other people's needs are never or only rarely considered. There are plenty of people around who believe that they are considerate of others, but they actually forget to ask about the needs of the other person in anything but a cursory way. Even then, the answer may not 'fit' with the original plan, and the other person's view is squashed because of their own fear of being seen as selfish.

Does this ring any bells with you? If anyone says you are selfish, the best thing you can do in finding out the truth and avoiding such a thought becoming another gremlin to be dealt with, is to look at *their* hidden agenda. If you normally consider the needs of others first, second and third before your own, it is essential that you improve your self-preservation skills by learning to say 'no' and putting your own health and personal needs at the top of the list in a more self-worthy position, particularly in times of difficulty or hardship (including illness).

Just to be crystal clear, control and manipulation of others is not part of the coaching ethic. What we are working on is enabling you to be capable, strong and self-reliant, with no need to resort to whining, pleading, emotionally blackmailing or any other guerrilla tactics that may be employed by people who are threatened by or frightened of life.

So, think about what goes into being a good friend. Write down all the characteristics and attributes that you feel you give to others in your friendships, and what your greatest friends give you. What is it about a best friend that is special? Could it be unconditional friendship that has kept you together through the worst times and the best? How have you supported each other? If you find these questions difficult, think about how you would wish a best friend to be; after all, not everyone has a best friend at all times in their life. But that is about to change.

Look at what you have written and ask yourself, 'Am I as good to myself as I am to others, or they are to me?' If the answer is 'no', think about what it is that you regularly do to yourself that is not particularly friendly, accepting or even plain nice.

If you constantly criticise yourself, tell yourself you are no good, give yourself a hard time about your standards, looks, time keeping, size, shape, ability, laziness, or anything else, it is time to stop. Listen to yourself and catch yourself giving yourself all this negativity. The question then is, 'Would I

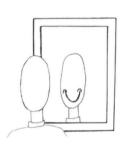

tolerate this unkindness from anyone else, and would I ever be this unkind to anyone else?' I bet the answer is 'no', and that you are much more forgiving and supportive of others. What makes you so different that you are more tolerant of the human characteristics of others than you are of

your own? Don't answer that! Just agree to be kinder to yourself, consider your needs, take time out for yourself and give yourself the same advice that you might be giving others. And enjoy doing so. The TV advertisements have a point – you are worth it!

ACTION POINT 3
My Personal Contract of Care

Another copy of this personal contract with yourself is at the back of the book (see Appendix of Charts) so that it can easily be removed, signed and placed somewhere visible in your home. You will not therefore need to write on the following page unless you choose to do so. You need to be able to see your contract everyday to remind yourself that you have made a commitment to doing the very best for yourself and your health. Signing it is an important step in deciding to look after yourself in a different way than you have done to date. As explained in 'Be Your Own Best Friend' (Success Skill 3), we are talking about keeping in mind all the strategies that will help you to move forward, including being kinder, more considerate and helpful to yourself in the way that you probably find it easier to be with other people.

My Personal Contract of Care

From today forward I promise to be my own best friend by
- ✓ Stating my health needs and wants in an adult, assertive way
- ✓ Caring about what happens to me
- ✓ Enlisting appropriate support that still enables me to be 'me' without losing control of my life.
- ✓ Giving myself praise for my efforts and achievements, however small I feel they are.
- ✓ Learning to say no when necessary for my own health and welfare

✓ Encouraging others to take their own personal responsibility.
✓ Taking responsibility for myself and my decisions.
✓ Believing I can make a difference to my life.
✓ Using the Flipover skill every day.
✓ Building rest, relaxation and activity into my daily routine.

Signed_____

Date_____

Review
In Chapter 1, we started to acknowledge your past efforts and began to 'Recognise and Zap Personal Gremlins' (Success Skill 1)

In this chapter, we have begun to choose a different way by:
o activating the Flipover skill (Success Skill 2) which can be applied to every negative situation you are faced with
o using some powerfully motivating questions to help you take control when faced with a problem
o putting into practice Gremlin Zapping (Success Skill 1) and the Flipover (Success Skill 2)
o thinking about and working on Being Your Own Best Friend (Success Skill 3)
o being ready to make and sign your own Personal Contract of Care, a copy of which is to be found at the back of the book (see Appendix of Charts)

Lots of positive thoughts and practical application. Well done! I hope you are making sure that you drink water at regular intervals during all these thought processes and that you are noticing when you need to have a rest from all this challenging thinking. Don't be surprised if you are tired. Just relax. The book will still  *be here tomorrow and, if you need to rest now, you'll be able to focus better at another time. There is no race or time limit involved here. You are in charge.*

3
Where Are You Now?

Every journey begins with one small step forward. Our ultimate goal is to for you to be living the life you want and successfully managing your Fibromyalgia. In order to move forward, we have to find out where you are now. There are a number of reasons for looking at this. It is useful to take a snapshot view of your present reality and then to do it again in a few weeks' or months' time. Then you will be able to recognise just how many changes you have made for the better. This can be a great boost to your morale, because although day-to-day changes may seem insignificant, over time they add up to a much greater and more obvious achievement.

It is also helpful to recognise how you came to be in your present circumstances, but don't worry there will be no long or drawn-out investigation with 'woe is me' attached to every thought. That would not be in the least bit helpful. Knowing and understanding about what happened to you does, however, help to support your new attitude to becoming well. Recognition of how you acted before enables you to adopt a fresh approach from now on and avoid the pitfalls of repeating the same scenario in anything that you are able to have control over. You can be sure that if you keep doing the same things then the results will be the same as usual, but, if you change even one small thing, then the result will be different. Any crafts people or cake makers reading this will recognise that changing a colour, material or ingredient will produce a completely different product. How does that thought fit with anything that you enjoy doing? From pigeon fancying to martial arts or football and all things in between, small changes bring different results, so why not apply it to your life getting through FM?

ACTION POINT 4
The Fibromyalgia Practicalities Wheel

This Action Point may look a bit daunting if you've never seen the like before. It is designed to help you think about all the practicalities of how your illness affects your ability to live in the way you really want to. It can be used to highlight your present needs and to help you recognise how much you are doing, alongside coping with your symptoms. In true 'Blue Peter' fashion, I have two completed wheels for you to see on pages 59 and 62. I hope this will give you an idea of how to use the results to your advantage. In Chapter 4, we also have an opportunity to work on more of the issues and find strategies for moving forward.

The wheel is divided into 8 sections. I have chosen 8 areas of life that I believe are vital for you to think about if you are going to give yourself the best chance of becoming well. Some are very practical issues, but others are linked to your inner being, strength, spiritual health and comfort. It will help you see what areas you need to look at first in your quest for regaining control over your life and living it to the full. It will also show how much you have already achieved in your journey back to health – we don't always notice the good things happening along the way.

How to use the wheel diagram
The sections of the wheel diagram reflect the issues that are particularly relevant at the time when you are coming to terms with the fact that Fibromyalgia is not a short-term condition.

In the real world, a wheel needs to be completely round to be effective and give a smooth ride. It will probably be no great surprise for you to see that your wheel, at present, is likely to be providing you with a very bumpy ride. You don't need this exercise to tell you that, but bear with me because it is part of knowing where you are, now, and where you are aiming to get to, from now on.

Take each section in turn, think about it for a little while then give yourself a score from 1 (which indicates you need to seek help) through to 10 (which shows that you are in total control and need no help in this area). Shade in that section so that you can easily see your present view of what life is like at the moment. Make sure you put the date at the top of the chart. The next time you do this exercise you will be able to see the changes you have made. Be honest. Pretending that you're okay when you're not will only delay getting things sorted out the way you would like them.

Here's an opportunity to practice 'Being Your Own Best Friend' (Success Skill 3).

✓ Get a pleasant drink or reward
  ready for yourself, for when
  you have completed your wheel.
✓ Be honest with yourself.
✓ Be kind to yourself.
✓ Know that you're worth it and
  you have the chance to work on it now.
✓ Take that chance.

The Fibromyalgia Practicalities Wheel

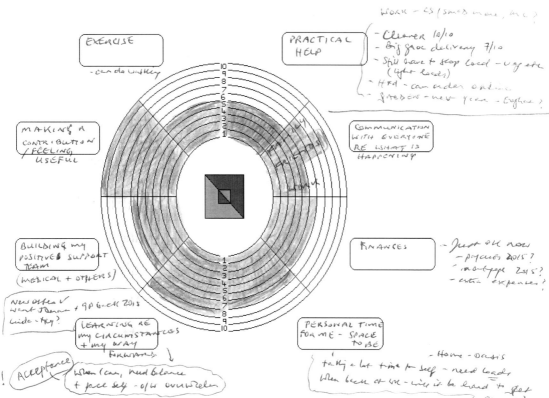

Here is the list of headings for each box with suggestions of how you might think about them as you mark up your wheel.

Practical Help: How much help do you have at present with cooking, cleaning, washing, personal care, shopping, caring for children and others, animals, gardening, and/or at work?

Communication with Friends, Family and Others: How much contact do you have? Think about the way you are able (or not) to be understood when you describe how you are feeling.

Financial Matters: How are you coping financially? This includes income or other monetary support, benefit, implications *of* illness, planning, where to get advice, debt management.

Personal Time/ Peace and Space for Me: How much time do you give yourself just 'to be' instead of 'doing' all the time? This includes meditation, relaxation, or any form of spiritual renewal.

Exercise and Healthy Image: How much do you really take care of yourself? Think about what you eat, how much you exercise, your clothes, and levels of self-image and self-esteem.

Learning about Me and My Circumstances: How much do you really know and understand about your illness? Do you know how to help yourself?

Building Positive Relationship with Medical Support Team: Consider all the conventional and/or complementary medical help you have. Are you an active partner in your own pathway back to health? Are you as factual and specific as you can be? Do you need to change your GP? What else do you need?

Making a contribution/Feeling useful: How do you feel? Working, paid or not, what can you still do even though your life has changed?

Using the wheel to prioritise your needs:
Whatever your wheel looks like, it is unlikely to show the contentment of 10 in each section right now. If it does, you have no need of this book! So, assuming that some sections scored lower than others, it is now up to you to decide which area is most important to deal with first.

Pat's wheel

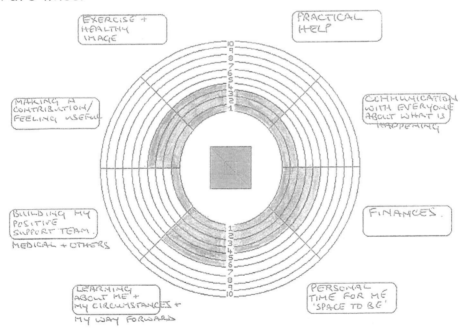

When Pat completed the wheel she felt she was hanging on to everything 'by the skin of her teeth'. Nothing was actually the way she wanted it and she had brought her home life into line in order to continue with her job, which she really loved. Even there, she was no longer able to ignore the reality of her Fibromyalgia. In reality, Pat went to work and returned home utterly drained and unable to do anything other than sleep. She was unable to initiate the things she wanted done in the house because of her feelings of reliance of other people, who were either too busy or doing their best in other ways to keep the family going. She considered her doctor less than supportive, so tended to avoid going to see him unless she was desperate. She was really quite depressed about her situation, and unhappy about feeling so negative when she wanted things to be much better and to get back to being the fun person she used to be rather than the 'nag' she felt she was turning into.

Over some months, during our coaching sessions together, Pat worked on recognising exactly what she wanted and worked out how to voice her opinion without feeling guilty, a nuisance or even a 'nag'. She gave herself time to improve her health by taking sick leave from work and working in a different way with her GP.   During that time she raised her self-esteem by searching out and learning about all the options she had for her working future. Originally she thought there were just two ways to do things – the right way or the wrong way - and she was extremely hard on herself in every area she thought about. She had never given herself permission to have fun because there were so many things she 'ought to do' first, but because of her illness and the pressure she felt, there was never any energy to do those things. The result was that fun was so far down the priority list it was never reached. Re-thinking those beliefs was the motivational lift that Pat needed. Suddenly, she felt better about tackling the 'must do' jobs. Having also realised that the whole job rarely needs to be done at one time, she began to pick away at lots of things and enjoyed seeing the differences she was making to herself and her surroundings.

Pat is now full of energy, but continues to be aware of the need to pace herself. She has returned to her beloved job, having set up over a number of months the exact situation that enables her FM symptoms to be minimised. As a valued employee who had learnt a lot about the disability at work legislation, she has helped her employers to be more effective in their handling of similar situations, and she is now working with others to encourage them along the same lines. She knows the pitfalls of being too conscientious and now remains totally professional, but within her personal limits of self-preservation. The last I heard, Pat was determined to keep the balance in her life and was enjoying laughing again.

Mia's wheel

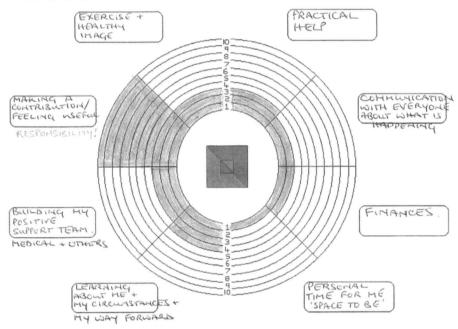

EXERCISE + HEALTHY IMAGE

PRACTICAL HELP

MAKING A CONTRIBUTION/ FEELING USEFUL
RESPONSIBILITY!

COMMUNICATION WITH EVERYONE ABOUT WHAT IS HAPPENING

BUILDING MY POSITIVE SUPPORT TEAM.
MEDICAL + OTHERS

FINANCES.

LEARNING ABOUT ME + MY CIRCUMSTANCES +
MY WAY FORWARD

PERSONAL TIME FOR ME 'SPACE TO BE'

Mia had no trouble feeling useful or knowing that she was making a contribution in her life. A truly capable lady, when she started coaching she was running two businesses and successfully holding down another job whilst trying, with difficulty, to manage her symptoms. Looking at her own wheel enabled Mia to see just how much responsibility she took and where she needed to make changes. In this case, the wheel exercise flagged up 10 not as an excellent but as something that was too out of balance with everything else. All the other issues that needed attention were just left. Quickly, Mia began to ask herself the right questions to enable herself to make the changes she needed. She regained her personal confidence in dealing with her own needs and began to make a manageable plan in small enough steps to help her feel that she was progressing on all fronts. Mia continues to make great progress in bringing her life into balance, which she has already found has made a positive difference to her symptoms and strength.

Now it's your turn.
Now that you have seen two completely different wheels and the way that they can be looked at, I hope you feel ready to fill in your own. Your reality is special to you, and someone else's story is not your story, although you may find some similarities at times.

You will find a copy of the wheel for you to complete at the back of the book.

A few words of warning.
- o Try not to get involved with any comparisons. Nobody's situation is worse than or better than someone else's because each person has a different tolerance and acceptance of what is right or wrong for them.
- o Nobody's journey is exactly the same as anyone else's. Fibromyalgia affects people in many different ways, so do not put yourself in a position where you are deciding how good or how bad you really are.
- o Having FM means that you deal with what you have to cope with. Although meeting and talking to others is often helpful, it can also be a bit disconcerting. My suggestion is to concentrate on your own needs and push forward using all your strength for your own improvement. There will be time to make decisions about

how you want to run your social life when you are feeling more in control.

It may seem that in this Chapter there are a lot of Action Points. Take heart! They are all to do with the wheel and how to use the information. I decided to split up the task into different action points so that you don't feel overloaded by what you have to do. Don't forget, take a rest, have a change of scene or do the next section when you are feeling up to it. Nothing has to be done in one sitting or in a rush.

ACTION POINT 5
Making a small change work in a big way

Action Point 5 is a step forward. You *can* do this. Keep it simple and choose to do something that is well within your capability at present.

Having completed your wheel, you are ready to start to make it work in your favour. It's time to look at small and easy ways to make a helpful difference to your present difficulties. As we progress together, you will be given more and more strategies, ideas and questions to ask yourself and apply to your life, so that, when you revisit the wheel, you will be able to notice just how far you have taken back control and moved forward.

Look at the wheel. Ask yourself what will take the least effort to change, but will make the greatest positive difference to the way you feel at present. It could be something like picking up the phone and talking to someone, making an appointment that you have been putting off, deciding to buy a gadget to help you manage in the kitchen or around the house. There is no right or wrong answer. Only you know what will help. When you have taken that first step, look again at the wheel and decide on making another small change. You're really on your way now!

Your lowest scored section is where you need to put in the greatest work, but *any* difference in your thinking at this time is change, so don't rush! All the skills and action points in this book can be applied, at any time, to the needs you have identified in your wheel, and it is important that the daily management of your health comes first, so keep up your spirits by realising that you are doing the best for yourself right now.

Review

In Chapter 1, we started to acknowledge your past efforts and began to 'Recognise and Zap Personal Gremlins' (Success Skill 1)

In Chapter 2, we activated the Flipover Skill (Success Skill 2) to enable you to turn negative to positive. We worked on some powerful strategies to help you take control when faced with a problem. You put into practice the two Success Skills you have already learnt and began working on Be Your Own Best Friend (Success Skill 3). You signed your own Personal Contract of Care.

In this Chapter, we looked at the real issues of illness and particularly Fibromyalgia. You completed the Fibromyalgia Practicalities Wheel and now have a snapshot view of your feelings about your life at the moment. You practised Being Your Own Best Friend (Success Skill 3) by deciding to be completely honest about your feelings and being kind to yourself at the same time. Having looked at other people's wheels, you now know how to make the best use of your wheel information and how to choose where to start!

Don't underestimate how much progress you are making. We are laying the foundations for you to take back the control that you are likely to feel you have lost since the onset of your illness. Remember that there is no expected speed of progress

- you do what you can when you can. Another thing to remember is that all these things need practice, so don't be worried if you forget or make mistakes and wish you had done things differently. Use the experience and decide to do it differently next time.

Making a small, simple change can very quickly make you feel better about something that has been bugging you for ages. Then you're likely to feel the adrenalin rush of achievement, no matter what the problem was. Keep going. If you're doing this regularly you will also be getting yourself into some sort of routine of taking care of your own needs. It all adds up. Well done!

4
The Fibromyalgia Reality Check

There are many different theories about how Fibromyalgia begins. Research also continues to try to identify who is most susceptible, how much a person's medical or personal history affects their chances of developing FM, and how different attitude responses to illness help or hinder progress towards wellness. As with any question, the answers are open to interpretation, but some facts are indisputable:

- o Fibromyalgia can start suddenly, following an accident or trauma
- o it can also 'creep up' over time and its beginnings only become recognised for what they were when the symptoms become too acute to ignore any longer
- o some people are coping with other debilitating illnesses alongside their FM
- o others have Fibromyalgia without any other recognised medical condition (other than the range of Central Sensitivity difficulties that often accompany FM)

Your FM, whether it is linked to other conditions or not, has already forced you to change. That change either crept up on you or hit you squarely in the face, seemingly without a great deal of warning. Whatever your own experience, the shock of that unplanned change and the knock-on effect on the way you now feel able to live your life, is bound to have caused you considerable stress, distress and perhaps even depression, which all have a huge detrimental effect on self-image and self-worth.

Have you noticed how difficult it is to explain exactly how you are affected by Fibromyalgia, or even how to describe what FM is all about? If, like many people, you look well and appear to have nothing wrong with you, it is easy to become depressed when friends and family find it hard to believe the reality of what you are coping with.

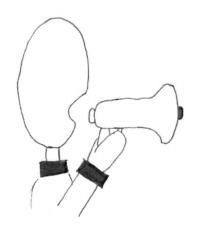

SUCCESS SKILL 4
Your Fibro-Speech: how you can help yourself and others by what you say.

Although all the Success Skills are important, Success Skill 4 will help you to be ready to deal with all sorts of difficult situations, which might otherwise have left you flustered or unable to answer because of frustration or other people's perceived critical tones, which normally arise out of their ignorance of the facts.

What is a Fibro-speech? It is a way of explaining factually and without emotion what Fibromyalgia is (in the simplest of terms) and how it affects your day-to-day life at present. Wherever you are in your personal journey with this condition, you need to review your Fibro-speech regularly so that you update the information and make it relevant to your present. As usual, my suggestions refer to anyone who is finding life very difficult at this time. Later on in your coaching progress (not necessarily by the time you have worked through this book though), when you have worked through and made positive changes, you will be able to add helpful phrases such as, 'I used to have difficulty with…, but now I'm fine with that'.

There are two parts to working on your Fibro-speech: deciding what to say and learning it by heart.

You may be surprised that I urge you to learn it by heart, but the very nature of FM can mean that when you find yourself in a stressful or difficult situation your mental powers of logical thought become disengaged from your ability to speak fluently. In other words, you dry up completely or waffle uncontrollably. I say 'you', but it's certainly what I was like, and my experience has been mirrored by most of my clients at times. Of course, there are plenty of people, without FM, who find speaking out under pressure difficult. However, when you feel ill, your confidence can be undermined by what is happening to you. When someone is apparently making you justify yourself, knowing how to respond by reciting your Fibro-speech enables you to feel in control of the situation and come away with your self-esteem intact.

Preparing a Fibro-speech and learning it by heart can help you deal with anyone who is interested, sceptical, thoughtless or critical. It also helps educate people about the condition, which is something that needs to be done at all times.

How you view your own difficulties also has an affect on anyone you meet. Here are two examples of a Fibro-speech. They are very different in their approach. Notice how you feel as you read them.

Fibro-Speech 1

"I have Fibromyalgia. I am in terrible pain all the time, and suffer from fatigue and exhaustion. I can't work, I can't write, I can't sit or stand for long, I can't do the garden, I can't cook, I can't clean, I can't shop. It hurts to drive, I can't concentrate for long, I drop things, forget things, I'm depressed and I don't sleep well. My life is no longer my own.

Fibro-Speech 2

"I have Fibromyalgia. It's a muscle-pain and fatigue condition, which is a hidden disability. It's not life-threatening but it is life-changing, and I'm learning about how to manage the symptoms and adapt. I have to do even the simplest things only when I can and only in short bursts, then take a rest. Some days are better than others and I'm learning to pace myself. It means that I have had to rethink even the most basic things about living to find ways round problems, but I'm learning a lot, and know that there are people out there to help me, so it's a challenge and I'm determined to live life to the full again, even if it takes some time to get there."

ACTION POINT 6
Choose your attitude now: it really matters

Which attitude do you think would be most helpful to you to adopt? What general affect on your spirit and your mental attitude does the word 'can't' have?

The second Fibro-speech acknowledges the problems in a more positive way: there is hope and determination, as well as recognition that life has changed and needs to be treated differently from now on.

If you are continuing to practise the Flipover (Success Skill 2) then I expect you are already thinking more positively about your situation than you were. Choosing your overall attitude to your illness is one of the most important decisions you can make. You may find it difficult to believe that life will get better for you, but it is partly your attitude that provides the key to progress. I have a feeling that if you are reading this book, asking the questions of yourself and working on my suggestions then you are already likely to have the right attitude. Don't give up on yourself though, even during the worst of days, just use Be Your Own Best Friend (Success Skill 3) and start again the next day.

An opportunity to use the Flipover (Success Skill 2)
Let's use the FLIPOVER skill with all the negatives in the first fibro-speech.

Negative	Positive
I'm in terrible pain all the time.	I have pain; its intensity varies.
I can't write.	I have to pace myself writing.
I can't sit for long.	I need to change position frequently.
I can't stand for long.	I need a chair to be available for me.
I can't do the garden.	I have to watch the grass grow at this time.
I can't cook.	Hooray for the microwave.
I can't clean.	I particularly need help to vacuum.
I can't shop.	I shop on good days. Hooray for internet shopping! I have great neighbours/ family/friends.
It hurts to drive.	I still drive on good days, but not far.
I can't concentrate for long.	I can concentrate for short spaces of

	time. It's better in the morning, afternoon or evening.
I drop things.	I have to concentrate on doing simple things
I can't lift things.	I need help with lifting because everything feels heavy.
I'm depressed.	I'm coping well in the circumstances, but I do get down sometimes, which is natural.
I don't sleep well.	I have to rest during the day at the moment.

Notice how there is still realism in the second, more positive approach. There is no minimising of the circumstances you are living through.

ACTION POINT 7
Design your own Fibro-speech

 As everyone's reality is different, it would be wrong to suggest that you take a pre-designed Fibro-speech and use it as your own. Take some time to work out what you want to say and what you can happily leave out, and make sure that you are giving yourself the support you need by choosing words that are factual, but not depressing in their tone. There is a very fine line between making everything sound totally negative, and being positive yet realistic about your symptoms and their effect.

Write it down, try it out on someone whose opinion you trust and revise it again if necessary. As you take greater control of your condition and manage the symptoms, you will find that your Fibro-speech will need to be revised even more positively.

JOAN needed her Fibro-speech to reflect her need to deal with people who expected her to be better every time they saw her (rather like someone recovering from a routine operation or a bad cold). She had got to the stage of only going out at times when she felt she would be able to avoid most people in the village where she lived. The difficulty in her communication was compounded by the fact that she and her husband had had a very active social life until her illness, and so many people were asking after her it had become almost embarrassing for him that there was no positive change and even more difficult for her as she had a time-related 'must get better' view of her condition.

Joan worked on what she felt were the most important points to get across to all the well-wishers who were unwittingly putting huge pressure on her when her sense of guilt and helplessness were already high. We came up with the following:

"I have Fibromyalgia, a pain and fatigue condition that is a hidden disability, therefore none of the symptoms can be seen. Although there is research going on, every person has a different experience of how it affects them. Unfortunately my symptoms are particularly severe at this time, but I know it is not going to kill me and I am doing everything I can to be well. I have to pace myself and every day is different. The good news is that, occasionally, I feel better which is a positive to build on."

When Joan originally completed her Fibromyalgia wheel, she marked the 'communication with friends and family' section only as 1, which reflected just how frustrated she was with trying to get people to understand what she was dealing with. Some weeks later she rated that same section as a 3, later a 6, and three months later she decided that her confidence in communicating well with other people had risen to 8. Joan had never really voiced her true feelings to outsiders, but the Fibro-speech (Success Skill 4) enabled her to speak up in the most challenging of moments and helped her to feel better about herself.

Review of what we have covered so far...
In Chapter 1, we started to acknowledge your past efforts and began to 'Recognise and Zap Personal Gremlins' (Success Skill 1)

In Chapter 2, we activated the Flipover Skill (Success Skill 2) to enable you to turn negative to positive. We worked on some powerful strategies to help you take control when faced with a problem. You put into practice the two Success Skills you had already learnt and began working on Success Skill 3 (Be Your Own Best Friend). You signed your own Personal Contract of Care.

In Chapter 3, we looked at the real issues of illness and particularly Fibromyalgia. You completed the Fibromyalgia Practicalities Wheel, giving you a snapshot view of your present reality. You looked at other people's wheels and now know how to make the best use of your wheel information and where to start. You practised Being Your Own Best Friend (Success Skill 3)

In this Chapter, we looked at how to help yourself and others by what you say. We investigated the power of the Fibro-Speech (Success Skill 4) before making a decision about your choice of attitude to the difficulties presented by Fibromyalgia. We put the Flipover (Success Skill 2) to good use in turning the totally negative into a more helpful, yet realistic positive. Finally, you took action to design your own personal Fibro-Speech (Success Skill 4), which will help you feel more in control when talking to others about your illness.

Lots of thought processes going on here! Beware the occasional gremlin of self-doubt creeping in. It is SO easy to feel that this is 'not worth it' or that 'it doesn't matter really'. What you are actually saying to yourself is 'I'm not worth it' and 'I don't matter'. Well, actually, you DO matter and what's more you are going to practise Success Skill 3 again (Be Your Own Best Friend), right now, by telling yourself that you matter. The only way is up, even if there are a few troughs and peaks along the path. Keep going, you can do it!

5
Keeping Records

In a society that is driven by the adage, 'time is money', it is not surprising that the need to 'make every second count' can backfire badly on such occasions when a truly listening ear is required in order to get to the root of the problem. Just when you long for someone to really listen to the difficulties you are experiencing without coming up with their own view too soon in the conversation, you are faced with a 5 or 10 minute doctor's appointment to get the whole picture across. For those of us who struggled on for a long time without bothering to see a doctor, the very act of setting foot inside the surgery is an admission of something being far from right. The modern day use of computer in consultations can also be a bit disconcerting if all you need is a bit of eye contact, empathy and understanding, in the first instance. One way to overcome this problem is to be very factual in your approach. Remember that not every doctor is well up on the problems of Fibromyalgia, and as there is no medication that is a 'one-fit' solution for everyone it helps to be able to notice any positive or adverse changes quickly if you are prescribed medication, and to be able to discuss the details as an equal partner in your own search for better health. You also need to be aware that FM is a condition that affects the mind, body and spirit; everything is involved in your wellbeing.

If you have a sympathetic doctor (and there are many of them around), there may still be a place for being able to give a clear account of the facts. Written records can save time and enable you to talk about the way forward more easily than a verbal description of what is happening to you.

SUCCESS SKILL 5
Keep To The Facts

The onset of Fibromyalgia
symptoms are difficult to pinpoint
unless they start as a direct result
of a sudden trauma, or immediately
following a virus infection. For many
people, intermittent aches and

pains, which seem to be manageable initially, encroach more
and more on their lifestyle. It takes longer to get up in the
morning; energy during the day is limited; you feel the need to
sleep during the day; muscles twitch, tighten and refuse to
stop hurting within a 'normal' amount of time. Dealing with the
balance between pain and fatigue, as well as coping with
everyday commitments, becomes more difficult. Muscle
weakness, pins and needles, numbness, sleep problems, and a
general feeling of being unwell creeps up on you until you are
forced to take notice of what is happening. To worry is normal
– who wouldn't be alarmed by the steady erosion of life as you
know it?

How to be believed: keeping records
The very nature of Fibromyalgia often means that when you
come through a particularly painful or difficult day the natural
tendency is to try to blot it out of your mind. Although this is
basically a very positive attitude, the down side can be that
you then forget the facts about how you were affected at the

time: as symptoms can change rapidly throughout the day and night, it is almost an impossible task to remember. But health professionals work on facts, not vague innuendo, so being specific in your reality description to a doctor is essential. A family member or any other enquirer will also be more able to accept visual evidence if they are truly interested. This type  of information, coupled with your Fibro-speech (Success Skill 4), can be the key to getting your message across and enabling people to understand your circumstances better. Feeling under pressure when trying to describe your problems can be stressful, and anxiety at such times may result in you underestimating or even forgetting some of the most significant things you need to explain. Coming away from a doctor knowing that you haven't said what you needed to say can leave you feeling deflated and frustrated with yourself. By keeping a record of symptoms and using it as an evidence tool to notice better times as well as difficult ones, you become more factual, focus on what is happening, and start to pace yourself more effectively.

ACTION POINT 8
The Symptom Chart

Symptoms are any feelings, problems or difficulties that give you cause to wonder about your health, or stop you doing what you would normally be able to do. Everyone deals with the occasional headache, tiredness or muscle pull, which can be easily remedied by taking medication or resting, but Fibromyalgia symptoms tend to occur randomly and may last only a short time but can also carry on for weeks, months or years with varying intensity. However, with a supportive team of people around you, it is possible to minimise the symptoms, so you may be relieved to know that I'm not asking you to fill in a symptom chart forever.

Once you begin to understand what sort of activity brings on certain symptoms you are in a much better position to be able to look at finding the right strategy to deal with them.

My suggestion is that you keep a chart going for two or three weeks in order to give you and your doctor a good idea of what is happening on a daily basis.

How to use the symptom chart

1. **Write down your symptoms in the left hand column.** Put everything down, including any bowel disturbance, fatigue, sensitivity to sound, light, food or touch. Include each area of the body that has muscle pain. You can always photocopy the sheet first, if you have more things to write down than there are spaces on the list.

2. **Record daily (morning, afternoon, evening and night) the strength or severity of each symptom from 0 – 10.** Make your own key to explain what the numbers mean. You could use colours instead of numbers if that appeals to you.

Here's a suggestion of how your key might reflect the numbers:

      0 would be 'no problem at all'
      1- 3 could be 'noticed but able to carry on'
      4- 6 'had to rest'

7 - 10 'unable to ignore' - had to stop completely and/or go to bed.

You choose what is best for you; after all, it's your chart and your body. If you feel that your description of the numbers reflects your reality you'll be able to explain it with greater confidence to other people. Again, there's no right or wrong way to do this; just make it your own and use it to help yourself.

3. **Mark any significant results,** for instance sudden peaks in pain or tiredness, and become aware of any triggers - reactions you have been having or activities you have been doing that may have a bearing on the symptoms. How you deal with this information is crucial to your wellbeing. Remember to use the Flipover skill (Success Skill 2) when assessing how much reaction or aggravation you get after doing something enjoyable; often the enjoyment is worth putting up with a bit of a reaction, especially as you learn to realise that you are not doing yourself any lasting damage and will come through the rough day or two once you put Success Skill 3 into operation by Being Your Own Best Friend again.

4. **Write down specific activities that you can manage, or have to stop, according to the strength or severity of the 0-10 scale.** Explain how this affects you emotionally too. Facts are not just about pain and fatigue, they can relate to the mental reaction that happens as a result of all that you are coping with. Any good doctor will be aware of the connection and give you sympathy and support when you need it. It's amazing

just how much a kind word of encouragement and understanding can help during difficult times.

Have you got your pleasant drink or treat ready to have when you have reached the point of needing a rest from this exercise?

If not, take a break and get it ready now and you'll be practising Being Your Own Best Friend (Success Skill 3) again.

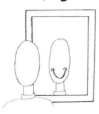

Symptom Chart

Name:

Symptom	morning	afternoon	evening	night	morning	afternoon	evening	night	morning	afternoon	evening	night	morning	afternoon	evening	night	morning	afternoon	evening	night	morning	afternoon	evening	night	morning	afternoon	evening	night

When BETTY (whom you have already heard about) completed her symptom chart in 2004 she added 'hurting all over', 'anxiety', 'light-headedness', 'burning/stinging', 'electrical impulses' and 'crawling feelings' as well as all the pain, numbness, tingling, pins and needles and fatigue. It soon became clear to her that the worst symptom for her to deal with was her anxiety about everything. The anxiety, of course, had a tension-like effect on her body that, in turn, produced more of the problems that were causing her anxiety. Somewhere, we had to break that cycle. We decided that the chart had done its job by highlighting what was happening on a daily basis, and Betty felt empowered to try to sort things out using all the options open to her. She discussed the way forward with her GP and started to look for ways to take action to calm down – easily said, but not so easily done, especially for such an active person who was used to being in control of herself. Together, we looked at all the ways that she could become more at peace without feeling guilty about taking time out from what she felt she 'should' be doing. It was important, too, that Betty should be able to lessen her anxiety levels without having to leave her house (having to be somewhere else to learn to be relaxed can often have the opposite effect) so she found meditation tapes, a video of Teach Yourself T'ai Chi and rediscovered her love of listening to music as she relaxed every afternoon. She also had reflexology on a regular basis, and of course continued with her coaching. Her anxiety about her Fibromyalgia lessened and became less obvious as she relaxed more effectively over the next few months.

Remember that you don't have to have ALL the symptoms mentioned, and you might even have different ones. There's such a variety of symptoms that you need to just focus on what you are dealing with, without comparing yourself to anyone else.

ACTION POINT 9
The Medication Chart: medication, supplements and complementary therapies

Action Point 9 is coupled with the symptom chart because it deals with what you are putting into or doing to your body in order to help yourself get better. We're not talking about food (that's a whole issue in itself) but prescribed or over-the-counter medication, food supplements, herbal treatments, and any complementary therapy that you have tried, or are trying, in order to be well.

Anything at all that you take or do to help yourself at this time should be documented. Your GP remains the first and constant port of call when things are not right with your health. Because FM is diagnosed by a process of elimination requiring all sorts of tests, this part of the journey can be particularly difficult. If you are trying complementary medicine or therapies, it is still important that you, at least, inform your GP of what is happening; better still, that you have a sense of equality with him or her and you both know you are working together to solve the problem.

Communication is vital, particularly if you have an unexpected reaction (good or bad) to some input. Too many people working on the same body at once can confuse the issue. Take responsibility for yourself by giving each new possibility a fair amount of time before deciding that it doesn't work. A word of warning however: if a drug is giving you side effects that seem to be worse than the original problem, do not waste time; get back to your doctor immediately to discuss a safe withdrawal.

Keep a record of any medication that you are taking. Although there is no magic pill, there are medications that work in some cases at certain times; for example, anti-depressants in a small dose are used to help with sleep patterns. A number of drugs prescribed for other conditions have also been found to help some people with Fibromyalgia. The difficulty is that as everyone is different, what works for one person may not work for someone else. This is a very good reason why you should take responsibility for monitoring your own response to any help. Remember that supplements and complementary therapies also impact on pain thresholds. Keeping a record of what treatment you have undergone and your body's reaction to that input is useful, particularly as the journey through Fibromyalgia is so very much one of trial and error.

How to use the medication chart
As with the symptom chart, use numbers or colours to make your own key to the way you fill it in. That will help you to

explain and to show visually what is happening to you without the need for anecdote or trying to remember feelings, dates and times under pressure. Although the medication chart looks suspiciously similar to the Symptom chart (which it is) the information that you provide on it can be used alongside the symptom chart to show up positive or negative changes in your condition. You are now becoming involved in what happens to you in a different way. No longer are you just hoping for someone to take control on your behalf, you have already made great strides to deal with what is happening to you.

Name:

Medication chart

Medication	Monday				Tuesday				Wednesday				Thursday				Friday				Saturday				Sunday			
Date	morn	off	eve	night	morn	off	eve	night	morn	off	eve	night	morn	off	eve	night	morn	off	eve	night	morn	off	eve	night	morn	off	eve	night

Review of what we have covered so far...
In Chapter 1, we started to acknowledge your past efforts and began to 'Recognise and Zap Personal Gremlins' (Success Skill 1)

In Chapter 2, we activated the Flipover Skill (Success Skill 2) to enable you to turn negative to positive. We worked on some powerful strategies to help you take control when faced with a problem. You put into practice the two Success Skills you have already learnt and began working on Success Skill 3 (Be Your Own Best Friend). You signed your own Personal Contract of Care.

In Chapter 3, we looked at the real issues of illness and particularly Fibromyalgia. You completed the Fibromyalgia Practicalities Wheel, giving you a snapshot view of your present reality. You looked at ways to make the best use of your wheel information and how to choose where to start. You practised Being Your Own Best Friend (Success Skill 3).

In Chapter 4, we investigated the power of the Fibro-speech (Success Skill 4) You then made a decision about your choice of attitude to the difficulties presented by Fibromyalgia. We put the Flipover (Success Skill 2) to good use in turning the totally negative into a more helpful yet realistic positive. Finally you took action to design your own personal Fibro-speech (Success Skill 4).

In this Chapter, we discovered how important it is to keep to the facts and record what is happening (even if you do this for only a short time) so that your reality becomes believable to those who need to know in order to be able to help. We looked at the Symptom chart, how to make it your own and how to interpret your findings as positively as possible. You had another chance to practice Being Your Own Best Friend (Success Skill 3) before looking at the Medication chart and how this can be used to monitor what is working and what may not be so good for you.

Many people resist completing forms mainly because they don't see the point or get bored with writing down the same things. These forms are about you and your Fibromyalgia. They are part of the strategy of enabling you to take control by understanding your reality and being able to show people, who

only deal in factual matters, just what is happening to you. Faced with the evidence of your detailed few weeks of records, any doctor, other health professional or occupational health officer will be able to see for themselves what you are going through. As a result of your hard work and diligence in completing this exercise, they are also likely to take very seriously your efforts to take control and get your life back on track. That will help you build your medical support team (see the Fibromyalgia Practicalities Wheel) and communicate with them as an equal partner on your road back to health.

6
What Does Change Mean to You?

Many people get anxious about change of any sort: jobs and responsibilities develop, personal relationships change, children grow up, and 'life just isn't the *same* as it used to be'. If you think of life as something that is done to you, or you feel that you are being swept along and have no chance to make any decisions for yourself, now is the time to think again.

We all make conscious or unconscious decisions many times a day. Each one has an effect on what follows, how we feel, and how others behave towards us.

Self-image, (which is made up of self-belief, self-worth, self-confidence and self-reliance) works from the inside out. If any of these elements are lacking, everything is affected and we function below what is possible for us as people. Is your self-image in need of a spring clean? Maybe it even needs digging up from some deep pit that it's buried in.

'Nothing wrong with *me*!' I hear some of you shout. Quite right! There's nothing *wrong* with you (in the sense of being right or wrong), but looking at how we *really* feel about ourselves, and reframing more positively how we *see* ourselves, is good for us all now and again. Ask yourself what's the worst that can happen by having a look? You might actually like what you find and be encouraged by that, or you might be a bit dismayed at how you've let yourself 'disappear' as a person, in recent times. For every one who says they are all right, there will be another five who know immediately that their self-esteem is low. The message is the same for all: have courage! With a bit of work, determination and commitment to yourself, your self-image will improve radically.

Although coaching is a future-oriented process, it remains important to find out where our beliefs about ourselves, and our values, have come from. Many people find it difficult to challenge ingrained conditioning about personal standards, life requirements, rules and expectations, which mostly stem from childhood. Some of our individual beliefs limit our view of what is possible for us in life. Lots of people are unaware that they have a right to hold their own individual views and opinions, and to be respected for those views, even if they are in direct opposition to those around them. It's never too late to change your own conditioning, to re-evaluate what is real about you and what is a myth, and to re-programme thoughts and feelings about yourself as a person. Self-image can be improved at any age and stage. And a healthy self-image makes coping with change much easier.

Another opportunity to practise Being Your Own Best Friend (Success Skill 3)

As someone else's best friend have you ever said, 'Come on, it's time you spoke up! I dare you to speak out – you can do it!'?

Now it's time to say that to yourself. Here are some ideas to help you on your way with this.

Your friends, family, workmates, and acquaintances see you as they believe you present yourself to the world. This may actually be quite different from how you think yourself to be, both on the outside, with social skills and behaviour, and on the inside, with your personal self-belief, self-esteem and self-confidence. If you live your life in many roles - husband, wife, partner, friend, colleague, parent, carer, brother, sister, aunt, uncle, granddad, nan, or workmate - remember that your contribution to other people's lives has been set up over time and that the change that Fibromyalgia has brought about in your life is also likely to rock the status quo in the lives of those around you. Some people may have an adverse reaction to the changes you have already been forced to make. This is usually because they do not understand what's happening (nor do you at this time!) and their own 'normal' or 'comfort zone' is threatened. They too, are uncertain of what is likely to happen in the future. The Success Skills that you have learnt already, in our work together, give you the tools to help yourself and everyone else to understand and support you in the way that is right for your particular family circumstances.

The impact that unexpected change can have on your life and the lives of those around you can be very daunting and rather frightening at times, too. But you are already doing everything possible to lessen the stress for yourself and others by looking at your situation from the positive coaching perspective, and you are adding extra skills and uncovering your own internal strengths as we go along. You may not be completely aware of this yet, but the more you practise the strategies and use the ideas we go through together, you will gradually see it more clearly and feel better about dealing with everything.

Coping positively with the fear of the unknown is one of the elements we will be working on as we move forward

together. For now, just remember that the only one you can really be sure you know what's happening with, and can make changes for, is *you*. Other people are likely to have a different agenda based on their own views, needs and wants. Unless you have young children (in which case you need extra support), you cannot be totally responsible for anyone at this time. This is the time to concentrate on you, and for everyone else to support and help you in the ways you have communicated (not as they want), or for them to seize their own opportunity to step up and grow up by becoming more self-reliant themselves. What a joy for you if some of your heaviness or worry is lifted as you recognise that everyone has stepped up a gear with self-confidence and self-reliance, and although they still love and want you around, they are no longer sapping your strength by appearing to expect you to take responsibility for them.

The next Action Point is designed to help you think about something you should, must or ought to off load from your 'to do' list in order to restore a healthier balance in your life. Normally I avoid using those words nowadays, preferring to think in terms of 'I would like to' or 'I choose to', which takes away the mental pressure of feeling that there is no way out of doing something. If you are person who values keeping your word once you have said 'yes' to someone, then it can be an enormous mental effort to say 'no' (unless you are in bed and too ill to move). Of course, such behaviour is taking the value of being responsible to the extreme. People begin to realise that you will always be there and they unwittingly (and sometimes purposely) take you for granted. This is where your understanding of limits should kick in, but because we are so used to being the backstop we lose sight of the limits and have to rethink what is happening to us.

GEMMA started the symptoms of Fibromyalgia following an accident. As a nurse, she was convinced that she had injured herself more than her GP and the specialist were prepared to admit. The true extent of her injuries (a broken clavicle) were not discovered for six years after the event, during which time she had constantly tried to tell her doctors that she felt something was really wrong with her shoulder, but she had never been truly assertive. In her own words, she had 'allowed herself to be fobbed off for far too long' and she was proved right. Because of her pain, she had had to give up nursing, and was subjected to a lot of medication and treatments. When she started coaching sessions, she was at the end of her tether, believing that everything was an uphill struggle and there was no way that anyone would ever support her in the way she really needed. She decided that she needed some different ways of looking at things.

    Gemma is a big-hearted, kind person who is always trying to lessen other people's loads. Whenever she reached her breaking point, her normal way of dealing with it was to 'blow a fuse' so that nobody could fail to know that she had had enough. At those times, most of her clan rallied round, but that see-saw behaviour only lasted as long as it took for Gemma to be back on an even keel again, physically, and then the family group went back to their normal expectations of her. In coaching, it didn't take Gemma long to see that she had to let go of the feeling that she was responsible for the wellbeing of everyone in her extended family, in spite of being really disabled and under very strong pain relief medication.

She identified some areas where she felt particularly stressed by being 'the fixer' for others. One particular area was her chickens: she is a hen rescuer and re-homes sick, injured or homeless hens; she takes them in, resuscitates them, cares for them and, when they are fit, takes them to new homes, which have been properly checked out first. She funds her activities by having coffee mornings and 'bring and buy' sales. She does all this, despite the severe health problems of FM, arthritic knees and sensitivity to certain medication. However, Gemma is no 'poor soul' and she would never see herself that way. She is a master of the 'Flipover', and her laugh, coupled with a wicked sense of humour, is completely infectious, no matter how grim she or you might be feeling at the time.

Gemma decided that some of her hen-keeping clients needed to be more active in the collection of new hens. Whereas she and her husband had always travelled miles to deliver hens to their new owners, she realised that this was part of being taken for granted, mostly because she had allowed it.

Quite soon, with her own health being put at the top of the agenda, Gemma reduced the number of hens she took in, which enabled her to have less of a daily responsibility towards the birds and their prospective owners. She also organised some help to clean out the cages and spoke up to a lady who had been in the habit of using Gemma's kitchen as a storeroom weeks in advance of the 'bring and buy' sales. Within a few weeks she felt less stressed in this particular area and began to apply the following questions to other areas of her life too.

ACTION POINT 10
Identifying your needs.

Give yourself time to look at these questions. I suggest you read them all the way through first, and then write down your ideas, starting from the beginning again. You could come up with huge issues or little niggling problems. Either way, it is important for you to see that you can deal with things in different ways. All the time you stay in the same groove, you will keep getting the same results. This is what I mean about speaking up for yourself and really thinking about  what you can let go of at this time. Believe me, although you may worry that the world will stop turning if you make some changes, it won't!

* What am I tolerating for no good reason?
* How could I lessen my load?
* With whom do I need to be honest?

- ❖ What is the smallest change that would make the greatest difference to me at present?
- ❖ What will happen if I don't delegate some responsibility?
- ❖ What will happen if I *do* make some changes?
- ❖ How will I go about this?
- ❖ When will I take action on making these changes?
- ❖ How will I feel once I have made these changes?

How do you feel now? A bit shocked or surprised at your answers? Don't worry, this is a bit like peeling the skins off an onion: the more you look, the more you find, but when added to other, new ingredients every bit of 'onion' is useful and helpful in contributing to a much more satisfying and enjoyable dish in the end. Have you ever thought of yourself as a potentially delicious, satisfying dish? What would you be?
When you look at your answers again, choose what to do first and when and how to do it, then just start picking away at the problem in small chunks, using all the skills you have already learnt. Hopefully, you are starting to feel empowered to make the changes you want.

I guess you are wondering, after that, just how much more action you will be asked to take in this chapter. Actually, it's going to be more fun than you might suppose. If you have shifted perspective in the way that  is possible by this stage in the coaching process, you may be happily surprised at what comes up in Action Point 11. It will give you a few minutes to rethink and write down other things about yourself that you may have forgotten, as well as to be kinder to yourself than you were during the first time you filled it in.

When EMILY first came for coaching she could only think of two things that she had achieved in all her 41 years: being married and having her son. In fact, she put them together as one achievement and then there was complete silence and consternation as she struggled to think of anything else she would consider an achievement in answer to the question. A few weeks later, we looked at the question again. This time, Emily reeled off a whole list of things, including the fact that from an early age she had been a singer in a band, loved performing and was a professionally trained musician who really yearned to get back to being able to make music again.

In Emily's own words, 'I am more than the amoeba-type person I seem to be at present; inside, the other me is screaming to get out.' Within three coaching sessions, she had made huge strides in understanding herself, and is now continuing to go from strength to strength in minimising her FM symptoms and getting her life back on track.

ACTION POINT 11
Reality check (just to be sure!)

THINK AGAIN ABOUT YOUR ANSWERS IN
ACTION POINT 1.

- ❖ Look at the way you answered the
  questions. How much did you belittle
  your efforts, or not give yourself
  enough credit for coming through the times you have
  mentioned?
- ❖ Be aware that you have already used inner strength,
  that you have dealt with a lot in the past and that you
  have come through to face yet another life challenge.
  Know that you can and you will get through this one. This
  time you will be armed with a self-knowledge toolkit of
  coping strategies and greater self-awareness.
- ❖ Revise your answers in a more positive light – use the
  Success Skills you have learnt and give yourself the
  praise you really deserve for how much you have coped
  with up to this point.

How do you feel now? Have you proved to yourself that you are
worth more than you gave yourself credit for the first time
you looked at those questions? If you are still not convinced
that you deserve any praise (especially if it's what you expect
of yourself, anyway), then ask yourself this:

- • what is it, *exactly,* about you that keeps you being
  so much harder on yourself than you are on
  anyone else who might be faced with similar
  difficulties

- why are you so unable to forgive your own mistakes and give yourself the credit you truly deserve for dealing with life's challenges
- are you really so terrible, or is it that you set the standards for yourself far higher than your expectations of anyone else
- what are you really trying to prove?

Being Your Own Best Friend (Success Skill 3) seems to be the most important skill to practice at the moment. So be kind to yourself, for once, and chill out! Think about how much of your energy goes into maintaining this 'super person' role. It's time to stop and join the rest of us in recognising that life is much easier once you understand that even the highest flyer, the most apparently organised person and those whom we admire from afar are still only human with all the same natural doubts and cares as everyone else.

It's ok to be human.

If you are still with me and taking your work towards good health as seriously as you have been forced to take your symptoms of FM, then you are doing a great job! Remember that everyone is different and that there will be good days

and bad days, as well as set backs that need time to get over, but you will never cover the exact same bit of road as before, because each time you come across an obstacle you are armed with greater understanding and more skills and strategies with which to cope.

Review of what we have covered so far...
In Chapter 1, we started to acknowledge your past efforts and began to 'Recognise and Zap Personal Gremlins' (Success Skill 1)

In Chapter 2, we activated the Flipover Skill (Success Skill 2) to enable you to turn negative to positive. You began working on Being Your Own Best Friend (Success Skill 3). You also signed your own Personal Contract of Care.

In Chapter 3, we looked at the real issues of illness and particularly Fibromyalgia. You completed the Fibromyalgia Practicalities Wheel and learnt how to use it to your best advantage.

In Chapter 4, we investigated the power of the Fibro-speech (Success Skill 4) and you designed your own personal Fibro-speech.

In Chapter 5, we discovered how important it can be to keep to the facts and record what is happening to you (even if you do this for only a short time). We looked at the Symptom chart and the Medication chart, and how to make them reflect your own circumstances in a factual, realistic, but also positive way.

In this Chapter, we looked at what change means to you and, as in earlier chapters, you have had lots of practice using the Success Skills you have learnt so far. You probably won't need a reminder of them but just in case...

Success skills 1 - 5

1. Noticing and zapping your gremlins
2. The Flipover
3. Be your own best friend
4. Your Fibro-speech
5. Keep to the facts

We completed this section by looking back and updating the original answers you gave in Action Point 1 when you first started.

Even the smallest positive change is worth a smile of satisfaction from you. It is not easy working alone, but I'm hoping that you feel supported by me as you make your way through these exercises and actions. Keep going! Every small step you are taking is a step towards getting your health back and enjoying life to the full. We're in this together.

7
Sorting Out Fact From Fiction

It is important for you to understand where your views about yourself have come from and whether they are in fact true. We are often our own strongest critics and it is the negative soundtrack in our heads that stops us from being the person we would really like to be. We see achievements in terms of 'yes, but...' or 'if only...' which waters down our true effort and sense of worth about what we have done in our lives, and also undermines the development of our self-belief. So many people get stuck in believing that there is a mapped path for all with gates to go through, and if you manage those you magically become a better person. There is *no* set path. Human beings are as diverse as any other creature on this earth; regardless of gender, race, origin, religious belief or sexual persuasion, we all choose to be the person we are. *You* make the decision about whether you wish to 'change the script' or are happy to keep it the same. *You* are in control of what you do. Personal change is not compulsory, but managing change with awareness and keeping some control over the process is easier on the mind, body and spirit in the long run.

Does this sound a bit scary to you? It could be that you have come to believe all the information that you have heard about yourself over the years. This could include things like 'none of our family has ever been any good at...' or ' you're just like your Great Aunt Doris, she was difficult too', through to 'you need to be ready for setbacks, life is like that'. The reality of accepting all this 'stuff' is that you begin to lessen your confidence and live the ideas. If people expect you to be like Great Aunt Doris, then you don't really want to disappoint them, so it's okay to be 'difficult' when in actual fact the reason you are having communication problems may have nothing at all to do with your own behaviour but plenty to do with the other person's problems.

MARGARET had always been told she 'would never amount to much'. The myth of her family was that they were all 'thick' and if she thought that she was capable of more then she was effectively stupid for thinking like that. A tough one to overcome and an apparently no-win situation. Home circumstances were not easy and yet ideas of wanting to do something worthwhile with her life wouldn't leave Margaret alone. Although she left school early and without qualifications her inner self-belief (she uses less complimentary descriptions such as pig-headedness amongst others to describe herself at that time) drove her on. She started training as a nurse, supporting herself with all sorts of jobs along the way. When she became qualified she still had to contend with back-biting comments from others who now considered her too 'posh' because of her professional status in a community where most people were not educated to a high level. However, those same people were extremely glad of her unconditional support whenever there was a health emergency or crisis and Margaret was on hand. Nobody said it was easy stepping out of the mould, but if you follow your true values with integrity (which some people call your gut instinct) you will find satisfaction, peace and enjoyment in what you do in the same was as Margaret did.

It has taken many years for Margaret to recognise just how strong she was in her youth and indeed still is now as she contends with Fibromyalgia alongside other health challenges. As others from her childhood neighbourhood also began to break the mould of family expectation she now recognises that she was the leader (albeit unconsciously at the time) and even her close family recognise her achievements more readily than ever before. It's never too late to change attitudes. It just takes one person to dare to be true to his or herself and keep going despite the odds!

STEVE's reality was different but he has had to work along the same lines as Gemma. Steve believed that every time he felt he was getting somewhere in life something else came to slap him back down. He had plenty of sympathetic friends who supported him in this view, and their conversations often turned into rants about apparently unchangeable things. Having left school without fulfilling all the potential he was showing academically, he chose a career path that was 'in the family' rather than looking at himself as an individual. When he gave up an engineering apprenticeship after spending 3 years struggling to keep up he went into retail and was successful in his work, but there was always somebody or something that gave him the feeling that he was being exploited. Steve had come through a number of real and devastating traumas in his twenties and early thirties which had affected him deeply, but Fibromyalgia was the last straw and he retreated into a sense of helplessness and hopelessness. The change came when he had to accept the fact that his physical and mental condition warranted the highest level of Disability Living Allowance award. It was not in his nature to be seen to be other than independent and he decided at that moment that he needed to use all his learnt knowledge about FM to work on himself with renewed effort in order to move forward.

Steve started using all the help that was on offer. He worked hard with a counsellor and was able to see why and how his own view of himself had contributed to all but the most serious problems he had encountered. He determined to keep working in slow manageable steps towards getting well. Recently he has found a new form of pain relief that works for him and he is beginning to feel better and discover exactly who he is and what he wants. He is becoming ever more strong and is now on his way to better health. He has built up a group of professional people around him who give him time and consideration, but also expect him to take responsibility for himself which he is now much more able to do. He believes he has more control over his own destiny now. The real Steve is beginning to emerge and the future is looking brighter. It's never too late to change!

ACTION POINT 12
What am I like? Recognising the real me

This Action Point is about sifting through the fictional information that you have collected during your life and deciding what is real and what is someone else's view, which you may have adopted as being the truth about yourself. Remember, nobody knows the real you except you. So approach this Action Point with a sense of 'I'm the only one who knows the true me' and you may be surprised at your answers. It also may enable you to feel freed up from some of the baggage you may have been carrying for years. I do hope so.

Using the chart below, write down all the things you have been told about yourself, throughout the years, that you have come to believe. Look at each sentence in turn. Who told you that? Is that *really* the truth, or is it another person's view and could they have been wrong, but you believed them? Discard anything that is not true. Be honest. Dare to look at yourself. Stand up for the real you, who is hidden deep down inside and has been covered by layers of 'should haves', 'must haves' and 'ought to's'! Sweep away the cobwebs that are other people's perception of who you are. Be confident in the fact that deep down, *only you* who you are, and what you believe yourself to be as a person. Put aside all thoughts of 'If only that hadn't happened I'd be...'

Things people say about me – positive and negative	True or not? Yes/no or sometimes	What am I *really* like as a person inside my heart and head?	Which bits of 'me' do I want to keep and which to change?

After you have completed this, look long and hard at the results, give some thought to the positive points you have written down and make a list of them in bright bold colours on a large piece of paper (or on the computer).

Look at your positive list every morning and evening, read it out loud and remember it so that you can bring it to mind if the gremlins of self-doubt pop up unexpectedly. Then you'll be practising Recognising and Zapping your Gremlins (Success Skill 1), which you may have thought we'd forgotten for a while. Our personal gremlins return unexpectedly and you need to remember that Success Skill 1 is one of the most empowering of all the Skills you have learnt so far. It has very little to do with Fibromyalgia and everything to do with the real you.

Remember, we choose, either to live our lives or to let others live them for us!

ACTION POINT 13
Making amends or setting the record straight, if necessary.

This Action Point follows on from your thoughts in Action Point 11. If it is not relevant for you, then give it a miss. However, giving some thought to how you can take back control of a difficult situation will help your self-esteem to rise. It is also helpful to admit that 'it takes two to tango' in all but the most serious scenarios. Of course I am not referring to anything that may have happened to you when you were a child, when adults had total control over what happened to you.

If the last few pieces of work have brought up any thoughts of anger, guilt, regret, or festering injustice within you, it may be a good thing to do a positive clear out before you attempt to move further forward. There are different ways to come to terms with something you are not proud of, or the feeling that you have suffered because of someone else's inappropriate behaviour toward you.

Here are some suggestions for you to think about.
- Do you need specialist counselling on this particular issue?
- If you feel you are at fault, would a simple apology help you to feel better? (Remember, that you can only be responsible for your own feelings, so be realistic in your expectations of how your apology may be received).
- Write a letter explaining how you feel, even if the recipient is no longer in your life. (They may even have died but it will still be worth doing if it helps you feel better.) According to the circumstances, you can send it, bin it or burn it. Whatever your choice, the result could be that you set yourself free of the negative feelings.
- If you feel anger, injustice or hurt, a letter to the person(s) involved may also help you to get rid of the frustration of keeping your feelings to yourself. It is up to you if you actually send it or not (if that's an option).

This is another opportunity to use SUCCESS SKILLS.
Maybe there are some personal gremlins here in which case,
you will be **Recognising and Zapping Your
Personal Gremlins** again (Success Skill 1)

Being Your Own Best Friend
(Success Skill 3)
and **Keeping to the Facts**
(Success Skill 5).

If you are getting really good at the **Flipover** (Success Skill 2),
you will also be able to recognise that some of the difficult

things that happened in your past have made you
the strong person you are today. That's why you
are able to take on all this work I am encouraging
you to do now. Well done and keep going!

SUCCESS SKILL 6
Think Outside the Box

Just when you think you are
getting to grips with 5 Success
Skills, along comes another one!
This one is the start of lifting
the lid off your thoughts about
your present situation.

Although it is not possible to wave away your Fibromyalgia
with a magic wand, it is possible to transport your mind away
from your physical problems in order to explore your thoughts
about yourself and your future.

If you knew you couldn't fail, what would you do, where would you live and how would your life be?
- ✓ Describe your thoughts in detail and write them down if possible.
- ✓ Live the reality in your head.
- ✓ Allow yourself to think without any negative thoughts spoiling your fun.
- ✓ Think about how you would feel.
- ✓ What would you look like?
- ✓ How would you interact with people around you?
- ✓ Who would they be?
- ✓ Where might you be?

This may be fiction at the moment, but it is freeing up your mind to be inventive and to search out exactly how you would like to be in the future. Humour me, please, especially if you think it's all a waste of time. It's always difficult seeing round the next bend in the road, but who's to say what great things are waiting for you there? Only you will find out on this journey, and by working with all these exercises you are giving yourself the best possible chance for your happiest future.

ACTION POINT 14
Start to think about what you really want. Dare to look beyond the present.

Action Point 14 brings you back to the present but in a different way from how we have looked at everything so far.

You are being asked to recognise what change means to you in a positive <u>and</u> negative sense; there are always two sides to a coin (which helps you also to remember to use the Flipover  (Success Skill 2) here).But we are also investigating and acknowledging all that you have already done recently, are doing and will do quite soon to help yourself. By completing this Action Point chart you are also practising all the Success Skills at once.

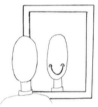

Coping with change

	Positive	
What change means to me......	Negative	
What I really want is......		
My reality today is...... Date:		
What I have already done to help myself......		
What I CAN do to make a positive improvement......		
What I WILL do this week......		
My promise to myself......		

When you have finished Action Point 14 have a good rest or take 'time out' in the best way for you at this moment. Have you listened to your favourite music lately? Or, like me, do you choose your music to reflect or lift your mood according to what you feel you need at the time? Either way, music or just silence is often forgotten in the rush of everyday life. If you are feeling energised by what you have done in this exercise, then do  *whatever you want but enjoy the feeling and don't worry if you need to rest more later.*

Review of what we have covered so far....
We're at the point now when we are beginning to look at things that may have held you back from feeling that you can do much to improve your situation with Fibromyalgia, or at anything else that you would like to change in your life. We are continuing to build on our work in previous chapters.
Here's a reminder of each chapter:
In Chapter 1, we started to acknowledge your past efforts and began to 'Recognise and Zap Personal Gremlins' (Success Skill 1).

In Chapter 2, we activated the Flipover Skill (Success Skill 2) to enable you to turn negative to positive. You began working on Being Your Own Best Friend (Success Skill 3). You also signed your own Personal Contract of Care.

In Chapter 3, we looked at the real issues of illness and particularly Fibromyalgia. You completed the Fibromyalgia Practicalities Wheel and learnt how to use it to your best advantage.

In Chapter 4, we investigated the power of the Fibro-speech (Success Skill 4) and designed your own personal Fibro-speech.

In Chapter 5, we discovered how important it is to Keep to the Facts (Success Skill 5) and keep records using the Symptom and Medication charts. We also worked on understanding how to use the information positively.

In Chapter 6, we looked at what change means to you and continued to practise using the 5 Success Skills you have learnt so far. We also did a review of how you are feeling now that you are half way through the book.

In this Chapter, we started sorting out fact from fiction to enable the real you to feel comfortable and safe enough to start to emerge. We set a few records straight and found strategies to deal with long held, self-limiting beliefs and we added 'Think Outside the Box' (Success Skill 6) to facilitate more forward thinking and planning.

Hopefully, this chapter really got you thinking in a different way from usual. Use the information you have found out about yourself, start walking tall, enjoy your new more positive mental soundtrack and start being the real you.

8
Getting Your Motivation Right

So far, you have had the opportunity to readjust your thoughts in order to be more positive about yourself, even though the reality of coping with your symptoms no doubt continues to be difficult and randomly disruptive to you and your plans. You have learnt 6 success skills up to this point and you need to use them on a daily basis in order to support yourself in the best way possible. It is really important to get back to the basics of who you really are (and to be confident with that), what you need and what you want for yourself now, and in the future. So many people spend so much time rushing around *doing* that they forget how to *be.* Using the power of the Flipover (Success Skill 2) at least you now have the time to make the changes that will be in your best interest for your improved long term health and happiness.

However, it's one thing being able to identify your needs and quite another having the energy and willpower to do something about them, particularly when the pain and fatigue of Fibromyalgia persists and it is difficult to concentrate. So, it's time to think about what really matters to you. What is it, exactly, about you and your view of life that motivates you to get moving, either to get up in the morning or to turn off the TV and get out there? What are you passionate about?

Understanding what makes you tick.
Every action and decision you have ever made has been motivated by something that matters to you. It could be an enjoyable feeling or emotion, or an inner sense of 'rightness' or justice for yourself or for others. Your decision-making process might also involve trying to *avoid* the onset of certain feelings or emotion. Guilt, anger, sorrow, fear and many other

human traits can affect the way we are motivated to take action, to put off things or to choose to do nothing. We all have a perception of what we see as positive or negative influences on our well being. Quite often this remains at a subconscious level, but working on understanding what makes you tick will enable you to be more proactive in your decision-making process. If you are a person who believes that your life is something that just happens to you instead of feeling that you can manage much of it, then this may be a 'light bulb moment' for you. I hope so.

Whose rules are you living by?
Your values can be more easily described as the rules you live your life by. Whatever your childhood was like, you grew up with a collection of values that came from the way the adults around you thought and behaved. Most people develop their own value system as they grow into independent adults, but very many more forget to off load the values that no longer serve them well. Some of these ideas are consciously or unconsciously buried and suppressed in the mind as the reality of day to day living takes hold. For many people, their 'rules' become wrapped up in how they see their roles in life. These roles can be empowering, but they can also be restricting if they are allowed to become more important than maintaining a healthy life balance and looking after yourself. Being an employee, employer, parent, child, uncle, aunt, friend, lover, workmate, colleague, carer, or any other role, becomes part of who we are and affects how we think we 'should' or 'ought' to or 'must' behave and react. We are usually our own harshest critics when it comes to giving time to what matters to us. This

is born out of our high expectations of ourselves, and also comes from the value system we have forged for ourselves along our life path. Add this to other people's expectations of us (just because we have always acted a certain way, people are used to accepting it as normal) and you have something that, in your present health difficulties, you need to address - pressure! In reality, it is *perceived* pressure brought about by being unable to see any different way round or through the problems.

One of CHRISTINE's most important values is 'family comes first'. She also values constancy and commitment. As a family-orientated Mum and Nan, her caring and coping strategies over the years had been to take on more and more responsibility and to try to be 'on call' for everyone at all times. Her strong sense of duty, which came from all the best feelings of family love, had developed out of her own childhood experience, which had not been quite so supportive. She perfected her supporting skills so successfully over the years that it was only her husband who really knew how often she felt overwhelmed, physically and mentally, by the effort it took to keep going, as well as to cope with FM and other long term health issues. Becoming increasingly ill, she felt more and more 'put upon' and became extremely tense, which, in turn, led to aggravation of her symptoms.

Christine had rarely refused anyone at any time. In fact, unless she was blinded by a migraine and couldn't drive her car, she would sometimes make superhuman efforts to support her children and grandchildren. In our coaching sessions we explored the fact that such dedication to the family may be having a detrimental effect not only on her health but also on the confidence levels of everyone else. Was it really that they needed her to be superwoman on their behalf, or did she feel that being superwoman was part of her role in life? What would happen if she wasn't there? Christine surprised herself by thinking of all sorts of other options and choices her family had in running their lives. Suddenly, she realised that this value of family could still be an important part in her life but in a different way. By doing too much for everyone, she had brought pressure on herself and she needed to give herself the permission to pull back a bit. She had, actually, encouraged her family to lean on her (possibly too much) and was now unable to continue in the same way. She also saw that as long as she continued in the role of being 'on call', the rest of the family would never experience the security of being totally aware that they can and will cope at times when she is not available. Christine decided to back off a bit in order to enable everyone to have the space to grow and develop in the way that she, actually, wanted for them.

Within a few weeks, Christine's family had begun to realise that Mum was not quite as robust and strong as she gave the impression of being. Having refused a few times when called upon to help in the normal way, her grownup children (who are all capable, responsible adults) quickly found other ways to work out their day to day family organisation, and Nan was less stressed and more fun when she did see every-one. Christine had quite a few issues to deal with during this transition – mostly guilt and taking the least line of resistance – which is not surprising after so long in such a major role, but she is now noticing the beneficial effect on her own feelings of stress, which have lessened significantly. She is also even more proud of her family.

Discard the unhelpful, embrace the helpful and enjoy being the real (or even the 'new') you.

ACTION POINT 15
Time for a values update.

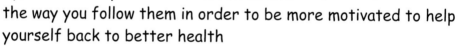

This Action Point enables you to look again at some of the 'rules' you are living by, to see how you can rethink them and the way you follow them in order to be more motivated to help yourself back to better health

What do you value? Write down the things that you have to have in your life in order to feel okay about yourself and your world. If you have difficulty being positive, think about what you hate or what makes you angry and use the Flipover (Success Skill 2) to *find* the positive side. For example, if you dislike people shouting and swearing, perhaps it is respect for others, good communication and peace that you value? Most people value money, but the underlying truth of that is more likely to be the security, the freedom or something else that having money brings, not the actual cash itself. Do you see how you look behind the apparently obvious to find the true value list? Take time to consider this carefully, and as you go about your daily routine notice what matters to you and how you feel when things are right for you. Also notice what is happening when you feel a build up of stress, tension or resentment within yourself. This will give you more clues about your values. Make a list of the ones you discover.

If you are tolerating things that are not in accord with your personal values, there will be a conflict between your present reality and what you feel is important for you. It's only when you are ready do more than merely say that you value your health that you will actually take positive steps on a regular basis to work on it. I remember feeling, initially, that nobody really understood the difficulties I was coping with. My extended family still saw me as looking all right, smiling most of the time and saying the right things to keep everyone happy. Some of my friends, who had been quite demanding in their needs, were draining me of the little energy I had. Of course, I take responsibility for letting that happen. Having asked myself, 'what am I tolerating for no good reason', I decided to decline the normal get-togethers, for a while, and gave the valid excuse of needing to focus on getting better, which meant using my energy in a different way. Immediately, a weight was lifted from my shoulders: my friends (who are really nice people) respected what I said, and I no longer felt pressured to conform. I had given myself permission to look out for me for the very first time! Did I feel selfish? No. I felt liberated, empowered and highly motivated to take my own quest for health seriously at last.

This is exactly why it is so vital that you give your values a 'spring clean', particularly now that you have had to make changes to your life because of your Fibromyalgia. Differences in value systems between partners, family and friends are the basis from which discontent grows. Not that everyone is expected to hold dear the same values (what a boring situation that would be), but understanding and accepting values for what they are, instead of believing that people are always annoying you on purpose, can be the greatest eye-opener to getting along better with friends and family.

You can only be responsible for changing yourself, not someone else. They are responsible for themselves, but when you make changes in your perception, and therefore in your reaction, you are taking ownership of your thoughts, feelings and actions, and the knock-on effect on others is usually dramatically improved.

Do you really value yourself? If you don't, then nobody else will either!
An interesting fact is that the profile of someone who develops Fibromyalgia is that of a caring perfectionist. You can be that in any walk of life and at any age. An overhaul of your life values can be the start of a completely different, healthier way of approaching the life you want to live.

It is our choice to work or live in a certain way. There will be those reading this who believe that they have no choice at all, but everything is choice. As I have said before, there is no one right way and no one wrong way of doing something. You only have to see disabled athletes in action to notice that. The  perfect body is not necessarily the fittest body. So if you have become stuck on the hamster wheel of life by believing you have no choice and that this is your lot, then think again. That's a challenging thought that I have mentioned before. It bears repeating because it is so important.

We all need breathing space, time for reflection and relaxation – time for ourselves. If you are one of the many thousands of people who have lost sight of life in the very act of trying to live it, it is a tough reality that unless you take a new look at your life, nothing is going to improve, particularly your health. The mind, body and spirit connection, or the lack of it, is a key element in the onset of such whole person illnesses as Fibromyalgia. Harsh words, you may think, but most people with chronic pain and fatigue, once they begin to think back, can clearly remember months or years of niggling aches, pains and 'worked-through' injuries that were dealt with the thought, 'I'll be fine, it's nothing.' You were not valuing yourself or your health at those times. What you were really saying was, 'I don't matter enough to put it at the top of my priority list.' The choices you make, *from now on,* will be more in tune with the needs of your holistic welfare than at any previous time in your life. After all, you want to be as well as you can be, don't you?

Set your goals according to your present values, and motivation will no longer be a problem.
Motivation is linked to your values. If you don't value something, you won't be motivated to take action, which is why it is essential at this stage that you readjust your value system to include good health. I can't make you feel that it's important, of course, but the big question is: if, because of your pain and fatigue, you can't do what you want to do with your life, where do you have to start to make a positive difference and take control? Yes, it's possible in many cases to invoke the disability legislation at your workplace, which will help you to remain employed, with support. But not everyone is in that position, and although this may form part of your successful management plan, if it is the only way you are approaching your FM management you are missing the holistic point. The Fibromyalgia reality is that it will not go away. My personal and professional experience proves that it can be successfully managed and its impact on life minimised effectively, but not without commitment and time from the person who is affected by the condition. So paying lip service to valuing good health is not enough.

Ask yourself what would having better health bring you. Now *that* could be a motivator and link back to you becoming really serious about doing all it takes to get well. Although personal circumstances vary, everyone *is* in a position to do the very best they can for themselves by paying attention to their own health needs, maybe for the first time.

We now have two Action Points in quick succession – just like buses, you don't see one for some time then two come along at once! But fear not, Action Point 16 is revisiting something you have already completed. Hopefully, this time you will be able to look at it with confidence and a new perspective. Action Point 17 uses the information available from your wheel and moves you forward with your small-steps plans. Here we go!

ACTION POINT 16
Revisit your Fibromyalgia Practicalities Wheel.

If it has been some time since you completed the wheel, the first thing to do is to check if any of your original scores have changed. Hopefully, they will be moving up the scale, unless, like Mia, your aim was to bring an unhelpful 10 down to a more balanced figure. Make sure that you colour in the wheel in a different tone to your original numbers and add today's date. An improvement in any area of even one notch is to be celebrated, so well done, and if it has remained the same, give yourself more time and don't worry. We haven't gone through all the strategies yet.

The next thing is to notice which areas have the lowest scores. These are the places to start making changes.

It is my guess that, once you have recognised that a small shift in some areas will make a big difference to how you are managing or feeling about yourself, you will begin to realise that you can apply the 'many small steps make one long journey' idea, and forgive yourself for not being able, at this time, to do everything yesterday and at great speed. If you are already recognising this – brilliant! You are definitely on your way forward! If you are still totally immersed in what you 'should' be able to do rather than what you can realistically do at this time, hang on in there, it'll come! All I am asking is that you make a conscious effort to try the strategies I am suggesting. Remember, it has taken a long time for you to reach your present reality. If you have become pessimistic about life in general, all the learned and assimilated negative thoughts and behaviour you have collected along the way are unlikely to be magically waved away overnight. Just as it took practice to perfect the way you think at present, so it is going to take practice to make the permanent changes in your thoughts that will enable you to be of greater support to yourself in your present circumstances. Remember to use Success Skills 2 and 3: Be Your Own Best Friend and Flipover the negative to a positive.

ACTION POINT 17
Make plans that respect and satisfy your values.

You have already identified your immediate needs from the 'Fibromyalgia reality wheel' and have made a list of priorities. Having worked, since then, on your inner knowledge of yourself and your life values, the next step is to work within this different framework when making plans for yourself.

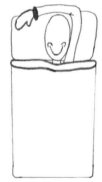

    The idea of setting goals can be a

negative activity for some people. If you have been beset with problems in your working life because of targets and goals, give some thought to the words you use and recognise that you are in total control of this process: goals and targets are no longer the immoveable objects they appeared to be when you had no control in setting them. Any successful plan has to be flexible enough to take on new aspects of development

along the way. You don't know from day to day (or even hour to hour) how your symptoms are going to behave or misbehave, so we are not adding a time element into anything we do here. If you decide you want to move forward on certain issues, your written plans and identified small steps can easily be changed if they become irrelevant to you. It is merely a system for making progress that reflects exactly what you see as necessary for you in order to satisfy your true values and be as comfortable as possible with your life whilst managing your Fibromyalgia.

Things change, feelings about things change, and just because you are writing down your plans, it does not mean that the words are written in tablets of stone, never to be revised as you think fit. Equally as important, however, is the well-known fact that people who make the effort to think things through and to plan their way forward by committing their ideas and plans to paper, usually get much further towards achieving what they have identified than those who just drift along hoping for the best. Action is the key to getting things moving, and we will deal with the pacing aspect of taking action later on.

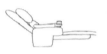

SUCCESS SKILL 7
Be in the Moment with Your Symptoms

Look at the goals or plans you have made for yourself. Are they too broad or expecting too much of you without taking into consideration the random effect of your FM symptoms?

This success skill will help you to cope with this without giving yourself extra stress. (And in the next chapter we will discuss in detail how to match your values with goals, as well as use a failsafe check before you begin.)

The intensity of your Fibromyalgia symptoms no doubt varies, not only from day to day, but also from minute to minute. Until you are able to notice, using the Symptom Chart, what your best time of day is and have recognised some patterns, you are at the mercy of how you feel *right now*. However much you want to get on with things, you are more likely to be able to do things better in the medium and long term by accepting that *at this moment* it may not be such a good idea. Instead of sliding into a pit of gloom, the words you need to say to yourself are, 'At the first moment I feel well enough, I will make a start.' Feeling well enough means feeling marginally better than you do at this moment; it does not mean waiting until you are feeling really well, which is likely to be unrealistic, at the moment. Do you notice how you are beginning to scrap expressions like 'should be able to', 'if only', and 'ought to'? These are very unhelpful in your present state so stick to accepting how you feel right now, saying 'no' to the internal pressure that may drive you to feel worse, and saying 'yes' to looking after yourself by remaining positive but still taking care of your present needs. After all, that particular feeling may have eased in a couple of hours' time and if you notice the difference you can do what you want to do later, or even another day, with greater success and satisfaction.

In the meantime, do what's best for you right now.
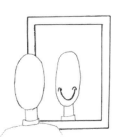
Some of the tried and tested ways to cope with pain and/or fatigue at such times are:
- ✓ give in, go with it and believe that this feeling will pass once you relax
- ✓ complete quiet and/or darkness or subdued light
- ✓ sleep
- ✓ gentle, slow stretching of affected muscles
- ✓ meditation
- ✓ self-hypnosis or relaxation tapes and CDs
- ✓ music
- ✓ warm bath with or without Epsom Salts, aromatic oils and fragrances
- ✓ wheat bag and/or hot water bottle
- ✓ lying on a padded recliner
- ✓ sitting or lying totally still with soft supportive cushions.

It may surprise some of you to know that some people actually force themselves to get out of the house and go for a walk in order to get rid of pain and/or fatigue. I also know of people who go for a swim and feel better afterwards. But once again, it depends very much on you as an individual. There are no hard and fast rules, so do what's right for you.

Some or all of these are coping strategies to try according to the severity of your pain and/or fatigue. Remember always to keep your body in a naturally straight line. The idea is to encourage and enable the muscles to relax; they will only do that if you make sure that they are placed where they should be, not scrunched up because you are miserable about the feelings you have. Keep going, you can do it!

Review of what we have covered so far. Hopefully, you are becoming familiar with all the Success Skills and using them at every opportunity.
In Chapter 1, we started to acknowledge your past efforts and began to 'Recognise and Zap Personal Gremlins' (Success Skill 1).

In Chapter 2, we activated the Flipover Skill (Success Skill 2) to enable you to turn negative to positive. You began working on Being Your Own Best Friend (Success Skill 3). You also signed your own Personal Contract of Care.

In Chapter 3, we looked at the real issues of illness, particularly Fibromyalgia. You completed the Fibromyalgia Practicalities Wheel and learnt how to use it to your best advantage.

In Chapter 4, we investigated the power of the Fibro-speech (Success Skill 4) and designed your own personal Fibro-speech.

In Chapter 5, we discovered how important it is to Keep to the Facts (Success Skill 5) and keep records, using the Symptom and Medication charts, and learnt how to use the information positively.

In Chapter 6 we looked at what change means to you and continued to practise using the 5 Success Skills you have learnt so far. We also did a review of how you are feeling, now that you are half way through the book.

In Chapter 7, we started sorting out fact from fiction to enable the real you to blossom. We set a few records straight and found strategies to deal with long held, self-limiting beliefs, and we added 'Think Outside the Box' (Success Skill 6) to facilitate more forward thinking and planning.

In this Chapter, we've not only identified the difficulties you are having (most of which only you would be aware of), but we have also considered your motivation and what makes you tick. We have looked at your values and where they came from. You have given them a spring clean so that you are now ready to start living according to what matters to you *now*, rather than how Great Aunt Doris (who may have long since joined the angels) would have expected you to live in their time. You have started to think about your goals and to match them to your values (there is more help on that in the next chapter) and finally, we have added 'Be in the Moment with Your Symptoms' (Success Skill 7), which enables you to take action at any time you feel motivated to do so, even if 5 minutes earlier you didn't feet like it.

We are really motoring with all this now, so if you find it all going a bit too quickly, just stop and go over some more familiar bits of the book again. It always helps to go through things more than once. As I explained before, small things can make a huge difference and you may find that some of the problems you felt were huge at the start of the book have now shrivelled into insignificance. Quite likely there are others that have shown up, but that's what this is all about and you can handle it, don't worry. Keep going!

9
Choosing the Right Way Forward for You

If you have read through from the beginning then you will already know that you have:

- ✓ found out what you really need to do to help yourself at this time
- ✓ investigated what really matters to you in life
- ✓ decided to be kinder to yourself
- ✓ started to know and approve of your inner self more
- ✓ become even more aware of your need to take your symptoms into account
- ✓ made your list of things you want to change
- ✓ begun to use the 7 Success Skills on a daily basis
- ✓ realised that good health is possible with small-steps action on all fronts.

As I explained in the previous chapter, we need to add one or two more strategies and skills to your ever-growing, self-help toolkit in order to make sure that failure is impossible for you. Just for a minute, however, I would like to encourage you to join the millions of people who have already been liberated from a terrible sense of failure, when they make mistakes, by using the Flipover (Success Skill 2) at such times. The question to ask yourself is: 'How or what can I learn from this?' Each time we make a mistake there is either something we need to avoid next time, or something positive we need to take from the experience (there's always a silver lining, however thin!) and put it to good advantage next time round. The word 'failure' conjures up the idea of the 'right or wrong' approach to life. This book takes a more balanced perspective. One person's failure may be someone else's triumph: it depends on their starting point and expectation of themselves and of others. I have heard many people say, 'I'm a failure because I didn't do this or that …' They forget that, in spite of that apparent failure, which may have been way back in their youth, they have actually achieved so much more in other ways. When they have this pointed out to them, usually the weight of that particular feeling is lifted from them for all time.

LINDA, who does not have Fibromyalgia, came to have coaching because she was torn between returning to her adopted country in the Southern Hemisphere and staying closer to her family in Britain. In order to get to know her thoughts about herself and her life so far, I asked her the questions that you have covered in Action Point 1 on page 33. Her answers were an unhappy account of how she had failed throughout her life. Seeing only a smart, confident, middle-aged woman in front of me, I asked when this sense of failure had started. Without hesitation, she answered, 'I failed the 11-plus.' (For those who are unfamiliar with this aspect of the British education system, the 11-plus was an exam for all eleven year old children, to determine their selection to various types of secondary school. Today, it only continues in certain small areas of the country.) Here was a 50-something lady of independent means still really upset by that 'failure'. I gently pointed out a few things about tests, selection and systems, which are never totally fair for everyone. I was also able to point out all the extraordinary successes she had had, in spite of this apparently traumatic failure. This lady had eventually been to University and was actually a highly qualified nurse with a long career behind her. However, this particular early 'failure' had pervaded every area of her life. Not only was it one of her gremlins that had turned into a huge monster (remember Success Skill 1), but she was still unable to recall it without tears, and had never considered 'zapping' it out of the way of her progress.

Happily, following some sessions with me she felt confident enough to make up her mind about where she wanted to be and, as far as I know, she is content with her decision. It is not only those with Fibromyalgia who have all these issues to deal with, and sometimes a sense of failure is almost as paralyzing as a really bad day's pain or fatigue.

SUCCESS SKILL 8
Look At Everything From All Angles

It's time to add another Success Skill to your list. It may be something that you are used to doing in your life anyway, but if you are unused to looking at things from a different perspective, then this is a good way to start taking a more considered view of events. So many of us rush in and make decisions, using only our eyes and our first reactions, and although intuition or gut reaction can often stand us in good stead, considering all the possibilities is more likely to bring about the happiest outcome.

Think over, under, round and through any difficulty. There are more ways to look at things than you may have believed until now. Success Skill 6 (Think outside the box) complements Success Skill 8 but they can both be used separately, according to need.

Here's a useful meditation, which can also be a practical exercise for finding a different view of any situation. Try it now, and let your creative thoughts take you wherever they want. If you feel inclined, you might even draw what comes to mind. Nobody is judging your artistic skill; if you fancy doing it that way, go ahead and enjoy yourself.

Imagine yourself looking at a cliff. Think of the detail - is it chalk, clay, grassy, steep, sloped, treeless or with lots of vegetation? Can you see or hear birds or animals? Where are you standing or sitting? Are you at the top of the cliff, looking down, or at the base, looking up, or somewhere else? Picture in your mind exactly what you see.
Now move in your mind to the opposite viewpoint. If you were looking up before, you will now be looking down from the top of the same cliff. What do you see and hear now? What is different about the view? Can you see all the same things you were aware of from your first position, or have some disappeared from view? Can you see extra things?

Now imagine yourself actually on the cliff halfway down. Are you clinging on to the cliff face or sitting comfortably on a ledge looking out at the horizon. How different again is your view of the cliff and the surroundings now?

We have only thought through three positions on that cliff. Even the smallest shift in position would provide a different insight into what the cliff is like. It would also give you a different sense of space and possibility.

Adapt this idea to your own thoughts about what is happening to you and what you want to do to make changes. Just 'go with the flow' for a while and be relaxed about it all. Once you stop expecting yourself to be full of ideas, your creative inner self will come to your aid. In fact, it's been there all the time; you've probably just been too stressed to hear it. If you think I am talking rubbish, that's okay with me because you are entitled to your view. Coaching is about mutual respect, too. I won't worry about it if you won't.

ACTION POINT 18
The values and goals match chart.

Although you have your list of things you want to change, and we have already looked at the idea of matching your goals to what you really care about in your life and the values you live by, you may find it easier to see everything by completing the chart in this Action Point. Once again, I urge you to make this chart your own by adding colour or any other information that will help you to move forward.

Before rushing in to change something, ask yourself the following questions and make the appropriate changes to your approach if necessary.

- By choosing this particular route, am I being true to myself and my values or am I still following someone else's value system?
- What will move me forward and make me feel good about myself?
- How can I change the way I tackle this challenge, so that it fits in with what I value?
- How can I look at this problem in a different way?
- What is best for the 'real' or 'new' me?
- Which of my values will I be satisfying by choosing to continue in this way?
- If I take this course of action/view, will I be respecting what I feel strongly about?
- Who in my life does not hold any of the same values as I do?
- What is the worst that will happen if I decide to spend less time with that person, or those people?
- How will I feel? Is that feeling likely to be better than I feel at the moment?

How to complete the values and goals match chart.
Look back at the work in Chapter 7 and 8. Don't be surprised if you have already changed some of your thoughts and priorities.

That's great! It means that you are moving and changing as you work. That's what we are aiming for, isn't it? Brilliant!

- ➤ List your values below box 1.
- ➤ List your areas of change below box 2.
- ➤ Take each area of change, decide on a realistic goal and list small steps in the boxes for each goal by asking, 'what is my first step, my next step, the next step …?'
- ➤ Be creative by using colour to identify which of your values is being satisfied in each of your goals and the small steps towards those goals.

VALUES AND GOALS MATCH CHART

1. My values:	2. Areas for change:	Goal + small steps towards	Goal + small steps towards	Goal + small steps towards	Goal + small steps towards	Goal + small steps towards	Goal + small steps towards

Just in case you are a bit overwhelmed by yet another chart, I am including one that I filled in myself to show you what it might look like. Remember your values may be very different from mine. And please, if you are finding it difficult to put one foot in front of another to walk anywhere at all, do not be intimidated by the fact that I am considering returning to a gym or doing other exercise that may seem 'off the planet' as far as you are concerned at this time. When I first started walking as part of an exercise regime, I approached the task with such over-enthusiastic determination that I pulled a calf muscle and ended up in bed for 4 days which was rather frightening, before deciding to try again at a slower speed. Then it was a very slow build up of stamina, always taking into consideration how the symptoms were each day. You know the sort of thing!

You are not therefore being encouraged to copy this, but just to see how it can be a good focus for recognising what your next small step is. If you look at each small step of progress I have identified and then look at the values I have listed, it is not too difficult to see those reflected in what I am trying to achieve. Health is paramount, and peace is gained through having all the other values satisfied – that's what I believe is right for me. If none of those values were being satisfied, trying to get fitter and having a holiday would be unrealistic goals to set for myself at this time. See how it works?

THE VALUES AND GOALS MATCH CHART

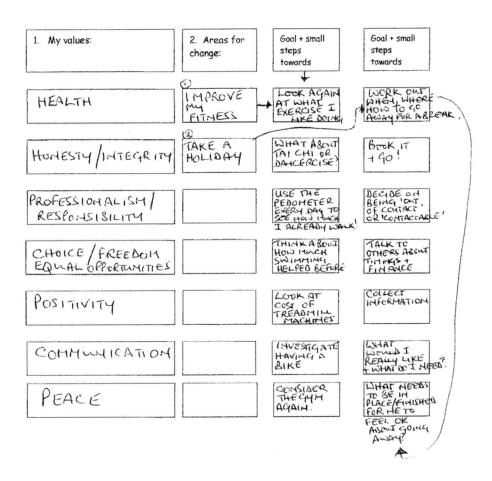

Life is constantly changing and all these strategies help to set up the most positive way forward, so this may be an exercise you return to every few months. I do. It may seem to some of you that this particular chart is in greater steps than you would be able to manage at the moment. There is a reason for that. I have been working on all these things since I was first diagnosed with Fibromyalgia in 2002. It is because I am so much better (and have been so for some time) that my planned steps are always much more than bite size nowadays. I still find it incredible to realise how far I have come myself, using all these skills, but in case you need further reassurance I'm including the details of a much earlier chart in the next chapter about pacing yourself.

You are definitely on your way now. Well done! Just *one more check*. This may seem tedious, but it really is necessary and will ensure that you always do the very best for yourself in the future.

SUCCESS SKILL 9
The Motivation Check

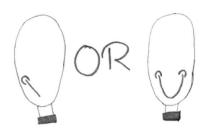

It's all very well me saying try this, have a go at that, and you feeling positive about things while you are reading in the quiet of your own company. But since the reality is that I'm not actually with you and your experience is exclusive to you alone, here is a final check to make sure that you have everything right for you. Here, honesty must rule! There has to be no messing about or fooling yourself. Truth is a must, or you are setting yourself up to do less than your best, which is not a bright idea after you have made such great efforts to move forward positively.

Ask yourself:
- On a scale of 1-10 (1 being not at all, and 10 being definitely yes), what is the REAL likelihood of my doing the thing I have said I want to do, by the time I said I would do it?

If the answer is less than 7 ask yourself
- What do I need to change or put in place first to make the answer a 10?

It could be that the timing needs to be changed, or that you need to get help, support or do something differently to make your motivation strong.

DAVID had already made huge changes to his life before he came to coaching and was certainly not frightened of change. When we talked about values, and matching goals to his values, it appeared that he was concerned about completing some small jobs that he had been putting off for some time, which had been a source of disagreement between himself and his partner of five years. He could not understand what it was that was stopping him from putting up a wall cupboard that had been sitting on the floor for months. When asked the question 'on a scale of 1-10 how likely is it that this task going to be completed', his answer was 2. In reality there was little or no chance that the cupboard would be fixed to the wall in the foreseeable future. David thought long and hard about the reasons behind his reluctance to take action. It came to light that his relationship was quite volatile, the house was not his, and his partner kept reminding him of that fact whenever they had a disagreement. No wonder he was procrastinating about the cupboard. When asked 'what needs to change to make this happen', David decided that he had to become more assertive in his relationship and establish a more favourable situation within their living arrangements. Either that,or to look again at becoming more independent, and to stop allowing himself to be threatened by words that might or might not have any bearing on his future. Within a few weeks David's demeanour had lightened considerably. He came in with a big smile, saying that he and his partner had worked out better ground rules for their communication. He had shared his true feelings and she had understood that her words had been unhelpful in the wider scheme of things. Within a couple of days the cupboard was on the wall!

Your procrastination may not involve such important changes to make before you free yourself up. It could be as simple as making sure …
   ✓ someone can accompany you for moral support
   ✓ you buy the right tool for the job
   ✓ you *really* want this to happen (things may have changed from the time you decided to make this goal)
   ✓ you have peace and quiet to enable this to happen.
   The motivation check (Success Skill 9) is extremely empowering and can be applied to every aspect of your life. Remember, the more you practise, the easier it becomes and the more successful you will become at sorting things out in a logical fashion that ensures you are satisfying your life values.

What we have covered so far:

In Chapter 1, we started to acknowledge your past efforts and began to 'Recognise and Zap Personal Gremlins' (Success Skill 1)

In Chapter 2, we activated the Flipover Skill (Success Skill 2) to enable you to turn negative to positive. You began working on Being Your Own Best Friend (Success Skill 3). You also signed your own Personal Contract of Care.

In Chapter 3, we looked at the real issues of illness and particularly Fibromyalgia. You completed the Fibromyalgia Practicalities Wheel and learnt how to use it to your best advantage.

In Chapter 4, we investigated the power of the Fibro-speech (Success Skill 4) and designed your own personal Fibro-speech.

In Chapter 5, we discovered how important it is to Keep to the Facts (Success Skill 5) and keep records using the Symptom and Medication charts and learnt how to use the information positively.

In Chapter 6, we looked at what change means to you and continued to practise using the 5 Success Skills you have learnt so far. We also did a review of how you are feeling about being half way through the book.

In Chapter 7, we started sorting out fact from fiction to enable the real you to blossom. We set up strategies to deal with long held, self-limiting beliefs and we added 'Think Outside the Box' (Success Skill 6) to facilitate more forward thinking and planning.

In Chapter 8, we found out what makes you tick and what motivates you to do something positive. We looked at values and started thinking about matching your goals to what really matters to you. We added 'Be in the Moment with Your Symptoms' (Success Skill 7), which helps you deal with positive as well as disappointing times without undue stress.

In this Chapter, we have worked in greater depth on values and goals. You have learnt to Look at Everything from all Angles (Success Skill 8) and have a useful perception shifting meditation to use. You also have a chart to match your goals with your values and a failsafe way of checking your motivation levels (Success Skill 9) and making changes in order to move forward with confidence and spirit.

10
Pacing Yourself

Studies of people who have been affected by the onset of Fibromyalgia have shown that the majority are busy, responsible achievers who have a caring disposition. If you recognise any of those characteristics in yourself, you may also be able to accept that making your own health and welfare a top priority has never crossed your mind before. Even if you have thought about it occasionally, you have probably found a hundred and one more important things that you 'should' or 'must' do before you get around to sorting out your own needs. It is reasonable to assume, therefore, that those who have always been very active find it totally shocking to be unable to function in their normal way. The fact that they are also known for being very responsible can impact greatly on those around them. It could be argued that, if you are such a person, you might have done yourself a disservice by being so organised in the past. Taking responsibility for all and sundry without regard for your own inner resources and needs can disable not only you but also those who need to learn to fulfil their own potential by taking responsibility for themselves.

Do not be disheartened by that comment. This may be the moment to stand back. Only you can decide to do that, and you begin by deciding how to proceed from now on.

Putting yourself at the top of your 'to do' list is likely to seem really strange, and may be quite difficult at first. Rest assured, it gets easier the more you practise, particularly as the benefits to your inner sense of well being and health become obvious. However, this change in attitude to yourself needs to be radical, and will remain fundamental to your own success in the management of your symptoms and future achievements, so don't underestimate its importance to your progress.

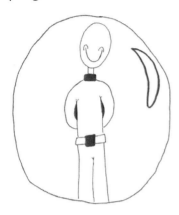

Self-preservation is not the same thing as selfishness. Human beings are entitled to be true to themselves and to live their lives according to their values and all that they hold dear. The reality in today's society is that the majority of people either have no time or no interest in finding out what is really important to them as individuals. They may have a sense of what others expect of them, but that is often far removed from what they would wish for themselves if they felt they had the freedom to reflect and make changes without fear of consequences.

Pacing, in the context of dealing with an illness, means using your energy and ability effectively, so that you do only as much as you are capable of doing at the time you want or need to do it. That could mean everyday things, but it could also be applied to special occasions and fun that will lift your mood, give you a sense of achievement, and provide a storehouse of memories to tap into at difficult times.

It is likely that you have rarely given much thought to how much energy you have been using without taking steps to replenish your supply. Taking care of your own basic needs enables you to do a better job in anything else that you are working on, whether it is within the family, out in the workplace, or both. Looking after your basic needs and pacing yourself will enable you to improve your own strength and ability to deal with your symptoms.

One of the most positive things you can do, now, is to start looking at your own situation as a project or a job in which you have the new opportunity to make all your personal, professional and caring skills work in your own best interest.

Identify the way you have worked in the past.
- Before FM slowed you down, how did you pace yourself?
- If you just kept going until you had to stop, what happened then?
- In any scenario, what was your motivation? *Deadlines? Contract? Work? Family commitments? Keeping everyone going? Not letting anyone down? Professional and/or personal integrity, pride, or even fear? Completing a job or fulfilling a role or a promise?*
- What made you feel really good? What still makes you feel good?
- If you have always pushed yourself hard, what is it that drives you?

As we found out in Chapter 9, matching goals to life values makes finding the motivation to carry them through much easier. Applying these thoughts to the subject of pacing will also help you to understand your own way forward more easily.

The way you paced yourself (or avoided the issue completely) before you were ill is likely to have stemmed from many different life experiences and coping skills. It might have been fuelled by enjoyment, fear, money, guilt, keeping the peace, making people happy, or anything else that motivates you. Identifying what has motivated you in this area is a step towards being able to pace yourself better now. It might even be a new experience for you to enjoy.

Did you ever make a decision based purely on your own physical, mental and spiritual needs, and not wrapped up in enabling you to complete something for someone else? Now there's a question!

I asked LIZ that question when it became obvious in our coaching conversations that she felt compelled to be the perfect housewife, in spite of having Fibromyalgia and a whole raft of other complicated health issues that had led her to give up work before she was 40. As a high-powered business woman, she had taken great pride in her management skills and, in spite of her chronic pain now, considered that it was her job to organise the home, cook and be the perfect homemaker in all respects. She held the view that 'anyone can run a home' and, from the viewpoint that at one time she had held down an exacting job as well as keep house, it is not surprising that she had envisaged it was going to be quite easy to do just the one job. The reality of coping with FM and her other health issues as well, was not quite so easy, although Liz rarely gave in to her true feelings of being overwhelmed by her own expectations of herself.

Her day started at 6 am with a rigorous pain management routine, then she tried to ignore her symptoms while she battled through all the things that she felt a perfect housewife should do. Her uncompromising viewpoint that 'anyone can keep house' gave vent to an inner competitive spirit that drove her forward and increasingly left her despairing of herself, because she actually couldn't cope with her own expectations. She held the view that anyone who sat down during the day was lazy, and when I asked about relaxation techniques, she actually used the word 'poppycock'. There was definitely some unpicking of unhelpful beliefs and values to be done, and also learning acceptance of the view that it's all right not to be perfect. Some of the techniques I asked Liz to take on were met with disbelief and a certain amount of scepticism, but I asked her to humour me, and she was so determined to be 'the model client' at that time that she agreed.

The change in Liz's attitude was gradual, at first, as we worked on freeing her of the idea that she had to do everything 'yesterday' or 'at breakneck speed' or 'all in one go'. We identified what sort of personal conversation she had with herself every minute of the day and, not surprisingly, it was critical and mostly negative. In fact, if she said aloud some of the phrases she was routinely saying to herself inwardly, she should have taken herself to a tribunal for bullying behaviour!

Pacing was the key to Liz to changing radically. She accepted that her job was now to become as well as she possibly could in order to live a balanced and happy life, rather than just a stressed one. Her partner of many years started noticing a difference in her attitude quite quickly and was initially startled, but delighted, by the difference in her response to events that previously would have created extra stress for her (and for him!). Liz, herself, was amazed at the difference her attitude had on others. She had expected friends and family to comment negatively, but they were all pleased and happy that she was now less stressed and able to say, 'I'll do that later,' instead of becoming upset that she hadn't completed the task immediately.

She talked about the 'new Liz' and the 'old Liz' to identify how she used to do things, and how it was now all right to behave differently. Within a few months of practising 'letting go' and pacing herself, without giving herself inner verbal grief every few minutes, she admitted that 'being able to pace without criticising myself is an absolute miracle'. She began to really enjoy not having the pressure of a job outside the home, and she said that she was 'drinking it all in, and can never remember being this relaxed'. Liz found that she was able to cut her medication down significantly, although her particular difficulties mean that she has to be aware of the need to continue monitoring and taking action as necessary. Liz is always busy with some project or other, but no longer feels the need to prove and justify her existence by being perfection personified. She is much more able to enjoy life.

Being kinder to herself and using pacing skills effectively enabled Liz to look more closely at what she really wanted. Over the years, she had put off wedding plans many times for various reasons, which were mostly wrapped up with being too busy trying to get everything perfect beforehand. Within a few months, however, the 'new' Liz decided that she was ready to be the wife she had actually always wanted to be, and she continues to enjoy her new freedom to be, as well as to use all the coaching strategies she initially thought were a bit silly.

So, what does 'pacing' really mean? In this context, it means going at a speed that is right for your reality now, rather than anything in your head that is telling you that you 'should' go at a faster speed. Just as an aside, when I first started yoga it never occurred to me that exercise could or would be better done slowly and with awareness. It took me a while to realise that being aware of my muscles and what I was doing was the best way of avoiding extra pain. That's always an incentive, isn't it?

ACTION POINT 19
Consider all options when pacing yourself.

Options are the foundations of choice. There is rarely no option: you choose the way to go after weighing up alternatives and consequences. Usually, we don't think widely enough when deciding what options we have, and often we forget that deciding not to decide right away is also an option, which helps us avoid making a decision too quickly and then regretting it.

Sometimes, your positive or negative thoughts about likely consequences are directly linked to your view of yourself or your situation. Often that belief is based on your perception and may not be based on true fact. It is important to find out the difference. (We worked on that in Chapter 7.)

Here are the questions to ask whenever pacing yourself is necessary.

- What do I need at this time? How can I make this happen for me?
- What action do I need to take? (Remember, doing nothing may be an option!)
- What do I really want?
- What can I let go?
- What is most important for me in this situation?
- What could I do?
- What will happen if I do this?
- What will not happen if I do this?
- How will be my health be affected by my choice?

Once you have given some honest attention to all these questions, pacing becomes easier. There are options at every turn. You *can* change and your life *will* be your own again; a different life from before, but nevertheless an interesting, worthwhile, balanced version based on your successful management of FM and what you really need and want to achieve. It *does* take practice to get out of the habit of regretting that you can't do things in the way you're used to. It also takes time, and you need to learn to be as committed to yourself and your improved health as you always have been to the projects, roles and caring you have worked on for other people.

Pacing is about choosing hour by hour minute by minute to work on becoming as well as possible. Avoid getting stuck in the 'I can't' mode. Remember that, sometimes, the aftermath of tiredness or extra pain is worth the effort of achieving something or going somewhere that deeply matters to you. Plan your days before the occasion; communicate with everyone to ensure you have rest facilities and appropriate cushions or seating available. Set your own parameters and stick to your plan no matter how much you may be urged to do more than you feel is in your best interest. There will come a time when pushing your boundaries will be the very best thing you can do for yourself and will help to show you just how far you've improved, but if you are, as yet, unused to being assertive and pacing yourself, stay safe in the knowledge that by so doing you are being creative and making good decisions for yourself. Make sure you feel totally responsible for you. If you allow yourself to be persuaded to move away from your planned arrangements, there should be no blaming of others; you have taken that option with full knowledge of what the consequences might be for you. That's the adult thing to do!

SUCCESS SKILL 10
Saying 'No' Without Guilt.

This Success Skill is the final one to learn. I left it until last because it is really one of the most difficult things for anyone to learn, but having already mastered the other 9 Success Skills you are in a stronger position now to take this one on with confidence. In fact, we are going to use all the Success Skills, together, to help you to say 'no' without guilt.

If you are already a person who copes with much, it is likely that you rarely refuse anyone anything. There is a saying: *If you want something done, ask a busy person.* But now is the time to learn to say 'no' from a position of empowerment for yourself and your health. The first thing you need to remember is that you have a right to refuse. Just because it is not your usual way of responding does not mean that the world will explode if you do something different.

We are going to invoke all the SUCCESS SKILLS as strategies to enable you to say 'no' without guilt or stress.

Looking after yourself from a point of self-preservation is not at all the same as being selfish.

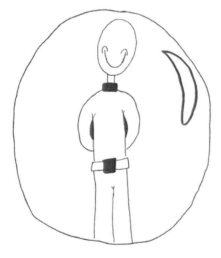

It is so easy to get into a mental spin in which every possible opportunity to move forward seems to be thwarted by the way others might or will react. Much of this can stem from what you have believed you had to tolerate in the past, and it also comes from your view of yourself and the role you have within family or other relationships.   Nothing has to be the same forever and, hopefully, you already believe that this is the time to change everything for the better.

Many books have been written about Self-Assertiveness (which has nothing to do with always getting your own way, but more about being an equal participant in negotiation and achieving a positive outcome in which everyone feels valued); therefore, there is no need to examine it in detail within the confines of this book. However, applying the Success Skills to your need to say 'no' will be an effective first step to becoming more assertive in everything you do.

The disappointment that can be felt when you believe that someone should have noticed something or could have helped more is no longer worth losing energy over. The fact is that people are not mind readers, nor are they all blessed with intuition or the sensitivity you may hope for. That is why open communication from you is extremely beneficial: firstly, to make the situation clearer; secondly, to help you to gain the appropriate support you need in your present circumstances.

Let's take, for example, a situation where you have 'gone along' with the decision to visit friends or family but you know that the reaction, mentally and or physically, is likely to set you back for weeks afterwards.

1. This is a gremlin of yours. How are you going to zap it? (Success Skill 1)
2. Flipover the negative to positive: what will you gain by not going? Or, what will you miss by going? (Success Skill 2)
3. Be your own best friend. What would you be telling someone else in your position? (Success Skill 3)

4. Use your Fibro-speech. Rehearse this factual information as your reason for saying 'no', or for making the circumstances more favourable for you to say 'yes', without anxiety. (Success Skill 4)

5. Do not be put off by questions. Keep repeating the facts. Avoid 'yes, but' and use 'the fact remains that I am...' (Success Skill 5)

6. If you really *want* to go to this gathering, think of new, helpful ways that will make things easier for you; for example, a bed available for a rest. (Success Skill 6)

7. Turn the invitation round by saying that you probably won't be there but if you are able to, you will be. (Only if you really want to be!) That enables you to be present in the moment with your symptoms and to stop yourself feeling extra pressure. (Success Skill 7)

8. When discussing the issue with family or friends, look at it from all angles but put your own health needs at the top of the list. If you keep doing what you've done in the past, then you'll keep getting the same results. Changing things will bring different results. (Success Skill 8)

9. Give yourself a quick motivation check. Is there something simple that could be arranged or changed that would make a huge difference to your confidence in your ability to be involved and not to have the usual negative reaction afterwards? (Success Skill 9)

10. All these thoughts can be applied to almost every scenario that you find difficulty in saying 'no' to. Saying 'no' enables other people to realise that you are serious in your intention to improve your health. It may also help others to start being more assertive. As

well as that, quite often other people stop taking you for granted and start helping in a more appropriate way. (Success Skill 10)

Again, this takes practice. If it doesn't work quite in the way you expect the first time, think about other ways to work at it next time, but use the Success Skills as a guide. You can do it!

'Where does the time go?' chart: Understanding how much you are doing every day

Having looked at saying 'no' without guilt, are you really aware of just how much you are doing in spite of having FM? The next chart is not really an Action Point, but if you are underestimating just how many things you pack into your day, you would be advised to complete it, if only to see for yourself and be proud, or to start saying 'no' to some of the unnecessary things you do through habit, duty or routine.

So many people completely undervalue their own daily tasks, activities and living skills that they get hung up on what they are *not* doing rather than what they *are* doing. If you are someone who thinks that no energy or mental effort goes into everyday living and you are disappointed that (because of your health issues) you seem to have nothing to show for your day, then the following chart is for you.

It can also be used to help you gather evidence for different agencies or your GP, but its most positive use is to show you how much you are achieving in spite of feeling ill.

Then, if you apply Be your Own Best Friend (Success Skill 3), you might even be able to say 'well done' to yourself, rather than feel fed up at not being able to do what you are used to doing.

Write down, in detail, the activities you do. They can be as basic as getting dressed, getting bathed or showered and can include looking after children, pets or other people, as well as hobbies, stretching exercises, watching TV, meeting friends, going to self-help groups and keeping doctor's appointments. Whatever you do during your day, add it to your list and then put a dot or a tick whenever it has been a factor in your day. You will probably end up with lots of ticks or dots in some sections and not so many in others, but remember: this is a form of self-recognition relating to how much you are doing in spite of your Fibromyalgia, not because of it.

Where does the time go?

Weekly chart   Start date:

Activities	Mon	Tues	Wed	Thurs	Fri	Sat	Sun

ACTION POINT 20
Small-steps goal setting

I purposely haven't talked much about goals and goal setting, although we did touch on it within the context of matching what you choose to do with your true values. Some people find that writing things down is too intimidating, especially when, because of illness, they have no way of knowing if they are going to be able to carry out the actions required to be successful. Success, as we have already learned, is all about perception. I consider any small step of achievement to be a success if it is helping you towards your ultimate goal of better health and successful management of your life with (or without) Fibromyalgia. Hopefully by now, you too are seeing the benefit in the small-steps approach to health.

The last chart is chopped up into small sections. That's exactly what you do with your goals. A goal can be anything – even as 'simple' as getting out of bed and getting dressed. Decide on something that you really want to achieve. Write that in one of the boxes at the top of the page. Then go down to the bottom of the page to the lowest box and write down the very first small step you need to take in order to help you towards achieving the goal at the top. Think about the next thing you have to do, write that in the next box up. Continue filling in the boxes with the smallest, most manageable steps that you can give yourself and, as you feel well enough and motivated enough to see them through, colour in the boxes to show yourself that you are making progress towards what you want to achieve. You may surprise yourself by how quickly you are able to complete the small tasks but, even if it takes a bit longer, you know in your heart that you are progressing, albeit slowly. That's important in maintaining your self-esteem and positivity.

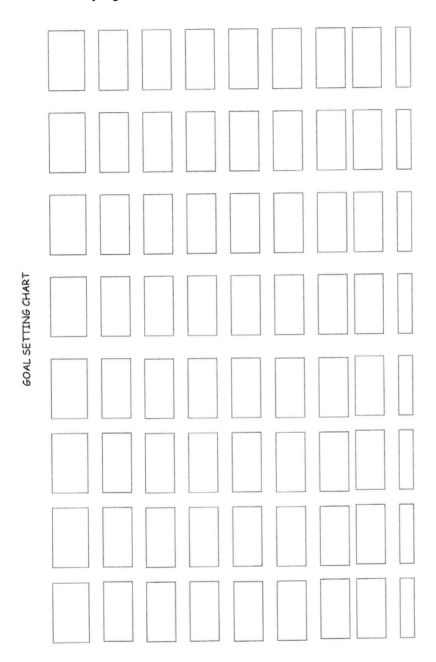

GOAL SETTING CHART

As I promised in the previous chapter, here is a copy of some of the small steps towards my own goals that I worked on in 2002. At that time, I put 'where I am now' in the small section at the bottom of each column, just to remind myself of my starting point. It was actually so difficult that I was highly motivated to move forward in whatever way was needed. If you look closely, I even questioned my own sanity and considered the possibility of self-sabotaging techniques. Going to see a counsellor and talking over the issues of hard work versus love of my job helped me to realise that it was only my physical problems that were keeping me from my beloved career. In some areas, people are more enlightened nowadays, but sadly I still hear of many extra difficulties being created, for those affected by such symptoms, by the lack of general education and understanding about Fibromyalgia and the way it can stop people in their tracks.

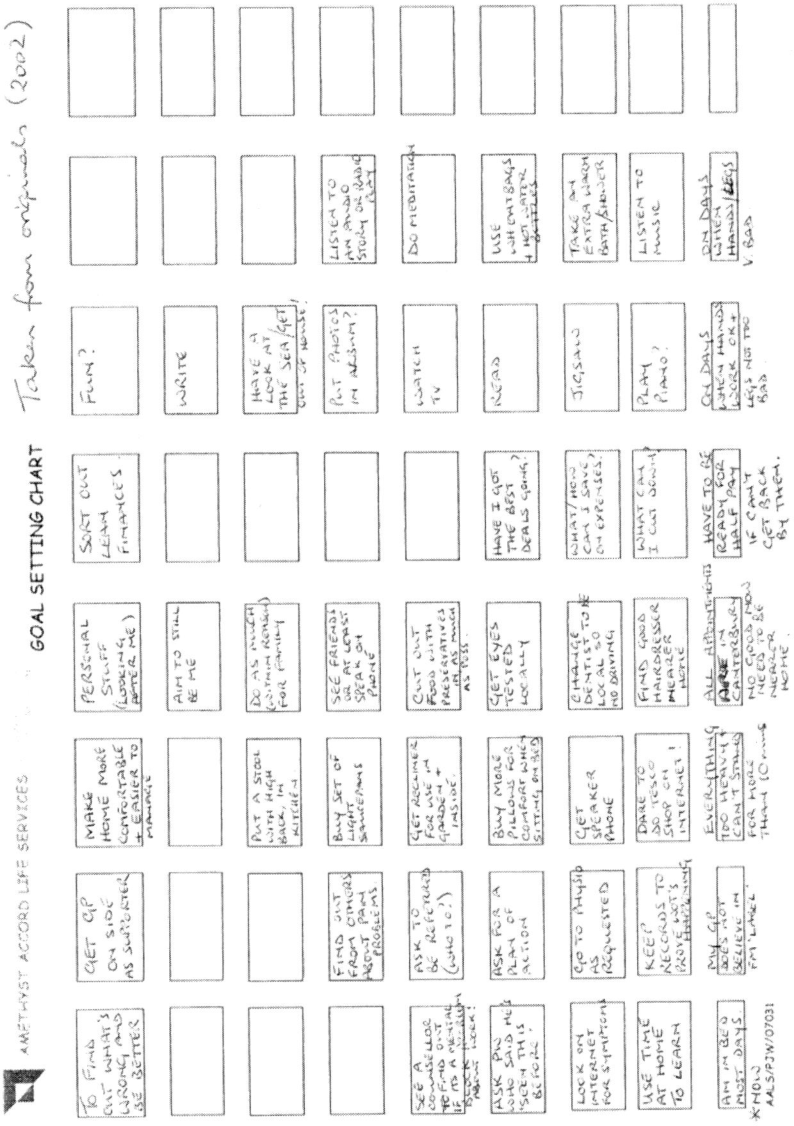

Taken from originals (2002)

AMETHYST ACCORD LIFE SERVICES   **GOAL SETTING CHART**

TO FIND OUT WHAT'S WRONG AND BE BETTER	GET G.P. ON SIDE AS SUPPORTER	MAKE HOME MORE COMFORTABLE + EASIER TO MANAGE	PERSONAL STUFF (LOOKING AFTER ME)	SORT OUT / LEARN FINANCES	FUN?		
			AIM TO STILL BE ME		WRITE		
PUT A STOOL WITH HIGH BACK IN KITCHEN		DO AS MUCH (WITHIN REASON) FOR FAMILY		HAVE A LOOK AT THE SEA/GET OUT OF HOUSE!			
FIND OUT FROM OTHERS ABOUT PAIN PROBLEMS	BUY SET OF LIGHT SAUCEPANS	SEE FRIENDS OR AT LEAST SPEAK ON PHONE		PUT PHOTOS IN ALBUMS?			
SEE A COUNSELLOR TO FIND OUT IF IT'S A MENTAL PROBLEM WITHIN — "PEOPLE THINK I MIGHT LOOK!"	ASK TO BE REFERRED (WHO TO?)	GET RECLINER FOR USE IN GARDEN + INSIDE	CUT OUT FOOD WITH PRESERVATIVES / EAT AS MUCH FRESH AS POSS				
ASK P.W. WHO SAID MLK "SEEN THIS BEFORE"	ASK FOR A PLAN OF ACTION	BUY MORE PILLOWS FOR COMFORT WHEN SITTING ON BED	GET EYES TESTED LOCALLY	HAVE I GOT THE BEST DEALS GOING?	WATCH TV		
LOOK ON INTERNET FOR SYMPTOMS	GO TO PHYSIO AS REQUESTED	GET SPEAKER PHONE	CHANGE DENTIST TO BE LOCAL AS NO DRIVING	WHAT/HOW CAN I SAVE ON EXPENSES	READ		
USE TIME AT HOME TO LEARN	KEEP RECORDS TO PROVE WHAT'S HAPPENING	DARE TO SHOP ON INTERNET!	FIND GOOD HAIRDRESSER NEARER HOME	WHAT CAN I CUT DOWN?	JIGSAW		
AM IN BED MOST DAYS	PAY G.P. DOES NOT BELIEVE IN FM LABEL	EVERYTHING TOO HEAVY, CAN'T STAND FOR MORE THAN 10 mins	ALL APPOINTMENTS ARE IN CANTERBURY — NO GOOD, NOW NEED TO BE NEARER HOME	HAVE TO BE READY FOR HALF PAY IF I CAN'T GET BACK BY THEN.	PLAY PIANO?		

*MOJ
AALS/PJW/07031

Extra column notes (right side):

LISTEN TO AN AUDIO STORY OR READ PLAY

DO MEDITATION

USE WM ONT BAGS + HOT WATER BOTTLES

TAKE AN EXTRA LUNCH BATH/SHOWER

LISTEN TO MUSIC

ON DAYS WHEN HANDS WORK O.K+ LEFT N.S T.B BAD

ON DAYS WHEN HANDS WORK O.K+ LEFT N.S T.B BAD V.BAD

What we have covered so far:

In Chapter 1, we started to acknowledge your past efforts and began to 'Recognise and Zap Personal Gremlins'. (Success Skill 1)

In Chapter 2, we activated the Flipover Skill (Success Skill 2) to enable you to turn negative to positive. You began working on Being Your Own Best Friend (Success Skill 3). You also signed your own Personal Contract of Care.

In Chapter 3, we looked at the real issues of illness and particularly Fibromyalgia. You completed the Fibromyalgia Practicalities Wheel and learned how to use it to your best advantage.

In Chapter 4, we investigated the power of the Fibro-speech (Success Skill 4) and designed your own personal Fibro-speech.

In Chapter 5, we discovered how important it is to Keep to the Facts (Success Skill 5) and keep records, using the Symptom and Medication charts, and learned how to use the information positively.

In Chapter 6, we looked at what change means to you and continued to practise using the 5 Success Skills you have learnt so far. We also did a review of how you are feeling about being half way through the book.

In Chapter 7, we started sorting out fact from fiction to enable the real you to blossom. We set up strategies to deal with long held, self-limiting beliefs and we added 'Think Outside the Box' (Success Skill 6) to facilitate more forward thinking and planning.

In Chapter 8, we found out what makes you tick and what motivates you to do something positive. We looked at values and started thinking about matching your goals to what really matters to you. We added 'Be in the Moment with Your Symptoms' (Success Skill 7), which helps you deal with positive as well as disappointing times without undue stress.

In Chapter 9, we worked in greater depth on values and goals. You added 'Look at Everything from All Angles' (Success Skill 8) to your self-help toolkit. You now have a chart to match your goals with your values and a failsafe way of Checking your Motivation Levels (Success Skill 9)

In this Chapter, we looked at options and choices involved with pacing yourself well. You completed your Success Skill collection by adding 'Saying 'No' Without Guilt' (Success Skill 10), as well as a chart to help you realise just how much you do every day, in spite of being unwell. You also have a small-steps goal setting chart to use.

11
Coaching is a Continuous Process

All the components for understanding yourself, thinking differently and using tried and tested skills, charts and actions are now in place for you to work with at your own speed and at a level that is right for your particular circumstances. This chapter gives you some insight into where we have already been on our journey together, as well as some suggestions for further general support.

Our journey so far

If you have read through everything, worked on the exercises and put into practice all the Success Skills, you will already have found that there have been some changes in the way you are thinking and managing your life now.

As I said at the beginning, nobody knows the reality of what you are living, other than you, but that is where you find your inner strength and begin to build on small steps of success towards what you really want and how you want to be. Once you unlock the real you, whether that is for the very first time, or whether you have released something that has been suppressed for some time, you will never return to where you were before. Recognising yourself and knowing what you are capable of gives you a perspective from which to face challenges with greater confidence and self-worth.

If you have followed the suggestions in this book, you have acknowledged how or why you reached the situation you are in at present. You have recognised that your negative gremlins are part of you, but it is how you deal with them that makes a difference to how you progress or become stuck. You have worked on changing your initially negative thoughts by using the Flipover skill to give a more positive spin on anything and everything, which can lighten even the darkest of circumstance - a smile (even a wry one!) is working your muscles and enables the 'happy' hormone, serotonin, to be released into the body.

Δ You have learned to be kinder to yourself by treating yourself as you would your own best friend, and you have started to speak up for yourself, factually and with authority, using your Fibro-speech and the factual approach to your symptoms.

Re-visit your Fibromyalgia Practicalities Wheel exercise and see how much has changed without you being truly aware of it happening; the evidence is there for you to work on. Use the idea again and again for things that matter to you, and make it your own by changing the headings according to what is happening to you at any given time. It is a useful tool with which to focus on what is really happening to you at any one time.

You have been encouraged to identify your true needs and to let go of inconsequential things that may have seemed important before but, because of your health, need re-evaluation.

We have covered some self-analysis, but all within the context of you becoming as comfortable as possible with yourself and your circumstances. It is this acceptance and self-understanding that will help you move forward in the right way for you. Remembering that Fibromyalgia affects and heightens *all* the senses (not just pain), some of this work may have been emotionally challenged, but it is nevertheless part of the way forward.

The motivation check and its link to your true values is a powerful tool, with which you can stop any procrastination and enjoy lifting the unseen blocks that you may have put in front of yourself when faced with an issue to solve.

Realising that there is no right or wrong way to get through this, but that everything is down to what and how you want to live, can be a bit disturbing but also very exciting. You have the tools now to do this with confidence. Pacing yourself is a long-term skill that requires thought and a certain amount of discipline. But if you slip off the path, just get back on again – this is not a points scoring exercise, it's real life and it matters.

Finally, you have all the charts to either use as you wish or to ignore until such time as you find a use for them. Some people love charts, others hate them. Knowing their use is only temporary, however, can give you a focus from which to spring forward more quickly. Make them your own and adapt them for your own needs. Again, there is no right or wrong way to address these issues. It is more important that you find them useful and fit them to your own circumstances and needs.

Options for support

There is an increasing awareness of the needs of people who have Fibromyalgia and other such invisible disabilities. Self-help groups are growing in number and there are many more on-line support forums. However, you must consider carefully the degree to which you get involved as all this takes time and energy that may be better placed in working on yourself. Some people really value meeting others within a group situation. A well run group can be a great source of information and camaraderie, but you should be aware that there are many different attitudes to long term illness, and you need to be able to disentangle yourself from any damaging negativity that you may encounter.

A few words of warning about time, energy and progress. Remember that nobody can put a time limit on your condition. Equally, nobody has the knowledge to say that you will or won't improve. Doctors are still learning about the condition and research is ongoing. It seems that amongst the greatest assets we have are the sharing of any news of success, and working in the way that suits each one of us best. It takes energy and determination to make the decision to be an equal partner (or even the leader) in your own improvement. As we have already discussed, progress can seem very slow when judged on a daily basis, but with constant commitment to yourself and your health you will begin to notice small changes, making you realise that it is possible to feel better. Even after a few minutes of experiencing that once, you know that the feeling is possible and that your hard work is going to be worthwhile.

As time goes on, and you have addressed all the issues and are working steadily towards your goals and looking after yourself in every area of your life, you will recognise the positive difference in yourself. Looking back over a few months, maybe even re-doing the Fibromyalgia wheel, you will see the evidence you need to know how far you have come. Hopefully, too, you will be enjoying life more, which in itself needs no chart to prove its case! Be prepared for setbacks or flare-ups, but rest assured that the severity of your initial symptoms is unlikely to return. Part of the coping strategy of a flare-up is not to panic; you know what it's all about and you need to keep using your strategies, perhaps taking even more care or rest while the flare-up lasts. This time, however, you've been there before and you know what to do to get better more quickly than when you were first ill.

SUPER SKILL

Knowing when to be independent, when to be dependent, and when to be inter-dependent, so that you are always doing the best for yourself, your health and those around you.

In this book, you have 'met' some of the people I have had the privilege to work with as they made changes to their own lives. Rarely do I meet people who have no longing to be independent again when illness has seemingly robbed them of the more obvious independent living that they were used to. It is a huge issue to deal with, but all the people mentioned learnt, subconsciously, and eventually consciously, when to maintain their independence fiercely (which was whenever they could), and when to accept help and depend on others for support.

Growing through such situations usually leads to an understanding of also being able to help others (verbally or in a practical way), in spite of it not being in quite the same way as it used to be. In other words, being ill does not stop everything you are or can be; it only slows it, according to the difficulties you experience with your symptoms.

Being inter-dependent means that you continue to hold yourself in high regard and offer help to others when they need it, whilst being comfortable with being independent whenever you can be or being dependent whenever you need to be, without guilt. This is what humans do, and this is the best way of describing 'give and take' without condition or manipulation. Idealism? I think not. Give it a go and see what happens? But as I suggested early on, do not look for brownie points and don't expect anyone to give something back. It's more likely to just happen then, without your controlling influence. Let the Universe work the way it loves, by providing what is needed at the time you need it.

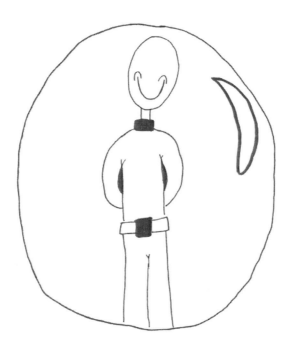

12
Keep in Touch

It has been a great privilege for me to be able to write this book in the hope of reaching a greater number of people with this practical, holistic coaching approach to living life well with Fibromyalgia.

This is probably just the beginning of the work on yourself. The more you practise the skills and action points, the more you will grow and make the changes required to enjoy life once again. Be ready to meet doubters who will not understand what you are doing. Some will be threatened by your changes, because even subconsciously they may gain something from having you stuck in one place. Just stick to what you want and need; hopefully, your greatest value is now your health. Nobody wants to be sick, but taking responsibility for all the parts of your life that you possibly can and looking after yourself from a position of self-knowledge is a great step towards feeling better about yourself, and can lead to greater things. Maybe your new life with Fibromyalgia will be twice as interesting, varied and balanced as the life you had to leave behind through illness. There are many thousands of people who successfully continue to work whilst managing their Fibromyalgia; it all depends on the symptoms, circumstances of employment and support available.

I would welcome your feedback. It would inform me of your future needs and help me in matching the content of future books to the ongoing reality of living life with Fibromyalgia.

Research to conquer Fibromyalgia continues. In the meantime, many more people are finding their way through, successfully, by using life balance and health coaching methods to take back control of their life.

Information about personal coaching is available on the website.

Single sessions or a series of coaching sessions are available whenever you feel you need specialist support.

Good luck! I look forward to hearing from you.

Pam

Website:
www.amethystaccordlifeservices.com
Email address:
pam@amethystaccordlifeservices.com

Appendix of charts

MY CONTRACT OF CARE

Sign and date this contract (or make up your own version) and put it where you can see it everyday.

From today forward I promise to be my own best friend by:

- ✓ Stating my health needs and wants in an adult, assertive way
- ✓ Caring about what happens to me
- ✓ Enlisting appropriate support that still enables me to be 'me' without losing control of my life.
- ✓ Giving myself praise for my efforts and achievements, however small I feel they are.
- ✓ Learning to say no when necessary for my own health and welfare
- ✓ Encouraging others to take their own personal responsibility.
- ✓ Taking responsibility for myself and my decisions.
- ✓ Believing I can make a difference to my life.
- ✓ Using the Flipover skill every day.
- ✓ Building rest, relaxation and activity into my daily routine.

Signed_____

Date_____

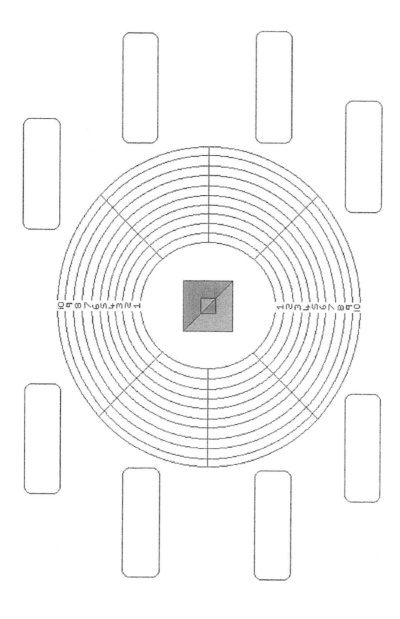

Symptom Chart

Name:

Symptom	morning	afternoon	evening	night	morning	afternoon	evening	night	morning	afternoon	evening	night	morning	afternoon	evening	night	morning	afternoon	evening	night	morning	afternoon	evening	night	morning	afternoon	evening	night

Medication chart

Name:

Date

	Monday	Tuesday	Wednesday	Thursday	Friday	Saturday	Sunday

Medication	morn	off	eve	night	morn	off	eve	night	morn	off	eve	night	morn	off	eve	night	morn	off	eve	night	morn	off	eve	night	morn	off	eve	night

Things people say about me – positive and negative	True or not? Yes/no or sometimes	What am I *really* like as a person inside my heart and head?	Which bits of 'me' do I want to keep and which to change?

Coping with change

	Positive						
	Negative						
What change means to me......							
What I really want is......							
My reality today is...... Date:							
What I have already done to help myself......							
What I CAN do to make a positive improvement......							
What I WILL do this week......							
My promise to myself......							

VALUES AND GOALS MATCH CHART

1. My values:	2. Areas for change:	Goal + small steps towards	Goal + small steps towards	Goal + small steps towards	Goal + small steps towards	Goal + small steps towards	Goal + small steps towards

Where does the time go?

Activities   Weekly chart   Start date:

Activities	Mon	Tues	Wed	Thurs	Fri	Sat	Sun

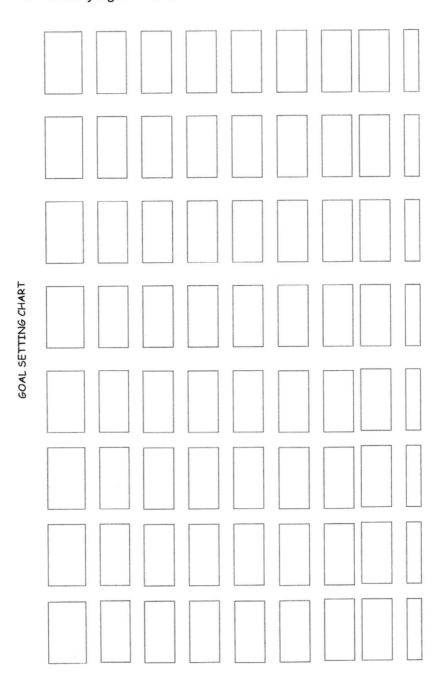

GOAL SETTING CHART

Acknowledgements

Grateful thanks to all those who kept me going when times were tough; a simple word of encouragement and unconditional love goes a very long way. To those who challenged me and made me rethink my whole life purpose, and to those whose unexpected support enabled me at last to believe in myself. Special thanks to my sons Tom and William Adnams, to readers of FaMily magazine, the 'girls' at Beyond Excellence Ramsgate 1 and Frances and Jill who appeared as if by magic to bring this book to you in its present form.